Michael S. Hildebrand, Gregory G. Noll,

# Aboveground Bulk Storage Tank Emergencies

SECOND EDITION

JONES & BARTLETT
LEARNING

*World Headquarters*
Jones & Bartlett Learning
5 Wall Street
Burlington, MA 01803
978-443-5000
info@jblearning.com
www.jblearning.com

Jones & Bartlett Learning books and products are available through most bookstores and online booksellers. To contact Jones & Bartlett Learning directly, call 800-832-0034, fax 978-443-8000, or visit our website, www.jblearning.com.

Substantial discounts on bulk quantities of Jones & Bartlett Learning publications are available to corporations, professional associations, and other qualified organizations. For details and specific discount information, contact the special sales department at Jones & Bartlett Learning via the above contact information or send an email to specialsales@jblearning.com.



**Production Credits**

General Manager, Executive Publisher: Kimberly Brophy
VP, Product Development and Executive Editor: Christine Emerton
Executive Editor: William Larkin
Vendor Manager: Nora Menzi
VP, Sales, Public Safety Group: Matthew Maniscalco
Director of Marketing Operations: Brian Rooney
VP, Manufacturing and Inventory Control: Therese Connell
Composition and Project Management: S4Carlisle Publishing Services
Cover Design: Kristin E. Parker
Rights & Media Specialist: Thais Miller
Media Development Editor: Shannon Sheehan
Cover Image: © Jones & Bartlett Learning
Printing and Binding: LSC Communications
Cover Printing: LSC Communications

**Library of Congress Cataloging-in-Publication Data.**
Names: Hildebrand, Michael S., author. | Noll, Gregory G., author. | Hand, William T., author.
Title: Aboveground bulk storage tank emergencies / by Michael S. Hildebrand, Gregory G. Noll, and William T. Hand.
Description: Second edition. | Burlington, MA : Jones & Bartlett Learning, [2019] | Includes bibliographical references.
Identifiers: LCCN 2017040237 | ISBN 9781284112771
Subjects: LCSH: Oil storage tanks--United States--Safety measures. | Petroleum--Storage--United States. | Flammable materials--United States--Safety measures.
Classification: LCC TP692.5 .H55 2019 | DDC 665.5/42--dc23 LC record available at https://lccn.loc.gov/2017040237

6048
Printed in the United States of America
21 20 19 18 17 10 9 8 7 6 5 4 3 2 1

Courtesy of William T. Hand.

## Notice

Petroleum storage tank firefighting is extremely dangerous; 70 responders have died or sustained serious injury and illness while attempting to mitigate storage tank incidents over the past 50 years. There is no possible way that this text can cover the full spectrum of problems and contingencies for dealing with every type of emergency. The user of this book is warned to exercise all necessary precautions when dealing with hazardous materials. Always assume a worst-case scenario and place personal safety first.

It is the intent of the authors that this text be a part of the user's formal training in the response to hazardous materials emergencies involving aboveground petroleum storage tanks. Even though this book is based on commonly used practices, references, laws, regulations, and consensus standards, it is not meant to set a standard of operations for any emergency response organization. The users are directed to develop their own standard operating procedures and follow all system, agency, or employer guidelines for handling hazardous materials. It is the user's sole responsibility to stay up to date with procedures, regulations, and product developments that may improve personal health and safety.

No amount of technical knowledge assembled at the scene of an emergency can guarantee a safe and effective response. As is the case with any hazardous materials emergency, the best way to ensure the safety of personnel working at the scene of an incident is to handle the emergency using the incident command system while following standard operating procedures under the oversight of the incident commander and a safety officer.

# Table of Contents

© Photos.com/Getty.


Acknowledgments .................................. vii
About the Authors .................................. viii
How to Use This Book .................................. ix

**CHAPTER 1** **Introduction and Overview of Aboveground Storage Tanks ....... 1**

Chapter Outline ..................... 1
Key Terms ..................... 1
Introduction ..................... 1
Target Audience ..................... 2
Overview of the Storage Tank Industry... 2
Summary ..................... 5
References and Suggested Readings .... 5

**CHAPTER 2** **Standards, Codes, and Regulations... 6**

Chapter Outline ..................... 6
Objectives ..................... 6
Key Terms ..................... 6
Introduction ..................... 6
Consensus Standards and Codes ....... 6
National Fire Protection Association..... 7
American Petroleum Institute .......... 8
- API-650: Welded Steel Storage Tanks for Oil Storage ......... 10
- API-620: Standard for Design and Construction of Large, Welded, Low Pressure Storage Tanks ... 11

Underwriters Laboratories Inc.® ....... 12
Steel Tank Institute ..................... 12
Regulations ..................... 15
- U.S. Environmental Protection Agency (EPA) ............... 15
- U.S. Occupational Safety and Health Administration (OSHA) ................... 15

Summary ..................... 15
References and Suggested Readings ... 16

**CHAPTER 3** **Tank Design and Construction Features ........................ 17**

Chapter Outline ..................... 17
Objectives ..................... 17
Key Terms ..................... 17
Introduction ..................... 18
Cone (Fixed) Roof Tanks ............... 18
- Design and Construction Features ................... 18
- Venting ..................... 18
- Flame Arresters ................. 20
- Weak Roof-to-Shell Seam ..................... 20

Open (External) Floating Roof Tanks ..................... 21
- Design and Construction Features ................... 21
- Floating Roof Types ............ 21
- Floating Roof Seals ............. 22
- Shunts and Lightning Protection .................. 23

Covered (Internal) Floating Roof Tanks ..................... 24
- Design and Construction Features ................... 24
- Floating Roof Types ............ 25

Low Pressure Storage Tanks .......... 26
- Design and Construction Features ................... 26

Horizontal Low Pressure Storage Tanks ..................... 27
- Design and Construction Features ................... 27
- Structural Supports ............ 27

Pressure Vessels ..................... 28
Product Transfer and Movement To/From Distribution Facilities ...... 28

Storage Tank and Distribution Facility Safety Features . . . . . . . . . . . . . . . . . .29
Facility Layout/Tank Spacing. . . . .29
Product Control Options and Considerations. . . . . . . . . . . . . .30
Ancillary Equipment . . . . . . . . . . . . . . . . .33
Piping. . . . . . . . . . . . . . . . . . . . . . .33
Piping Supports . . . . . . . . . . . . . . .34
Pumps . . . . . . . . . . . . . . . . . . . . . .34
Instrumentation . . . . . . . . . . . . . . .34
Valves, Connections, and Couplings . . . . . . . . . . . . . . . . . .35
Fire Protection Systems . . . . . . . . .35
Loading Racks and Transfer Operations . . . . . . . . . . . . . . . . .36
Summary. . . . . . . . . . . . . . . . . . . . . . . . .38
References . . . . . . . . . . . . . . . . . . . . . . .38

**CHAPTER 4 Incident Management and Response Considerations . . . . . . . .39**

Chapter Outline . . . . . . . . . . . . . . . . . . . .39
Objectives . . . . . . . . . . . . . . . . . . . . . . . .39
Key Terms . . . . . . . . . . . . . . . . . . . . . . . .39
Introduction . . . . . . . . . . . . . . . . . . . . . . .40
Managing the Incident: The Players . . . .40
The Players . . . . . . . . . . . . . . . . . . .41
Managing the Incident: The Incident Command System . . . . . . . . . . . . . . . .42
Managing the Incident: Command Considerations. . . . . . . . . . . . . . . . . .45
Developing the Incident Action Plan. . . .47
Size-Up and Logistical Considerations. . . . . . . . . . . . . .48
Initial Planning and Size-Up. . . . . .48
Tactical Decision-Making Framework: The Eight Step Process© . . . . . . . . . . .52
Step 1: Site Management and Control . . . . . . . . . . . . . . . . . . . .53
Step 2: Identify the Problem . . . . .54
Step 3: Hazard Assessment and Risk Evaluation . . . . . . . . . .54
Step 4: Select the Personal Protective Clothing and Equipment . . . . . . . . . . . . . . . . .55
Step 5: Information Management and Resource Coordination . . .56
Step 6: Implement Response Objectives. . . . . . . . . . . . . . . . . .56
Step 7: Decontamination and Cleanup Operations. . . . . . . . . .58
Step 8: Terminate the Incident . . .59
Summary. . . . . . . . . . . . . . . . . . . . . . . . .59
References and Suggested Readings . . .60

**CHAPTER 5 Firefighting Foam, Water Supply, and Fire Protection Requirements . . . . . 61**

Chapter Outline . . . . . . . . . . . . . . . . . . . .61
Objectives . . . . . . . . . . . . . . . . . . . . . . . .61
Key Terms . . . . . . . . . . . . . . . . . . . . . . . .62
Introduction . . . . . . . . . . . . . . . . . . . . . . .62
Foam History. . . . . . . . . . . . . . . . . . . . . .62
Class B Firefighting Foams. . . . . . . . . . . .63
Selecting a Class B Foam . . . . . . . .64
Storage Tank Safety and Fire Protection Systems . . . . . . . . . . . . . . .68
Fixed and Semifixed Fire Protection Systems . . . . . . . . . .68
Foam Chambers . . . . . . . . . . . . . . .69
Foam Discharge on Open-Top Floating Roof . . . . . . . . . . . . . .70
Subsurface Injection Systems . . . .70
Semisubsurface Injection Systems . . . . . . . . . . . . . . . . . .71
Fixed Foam Monitors . . . . . . . . . . .71
Fixed Foam Handline Systems. . . .72
Fixed Low-Level Foam Discharge Outlets . . . . . . . . . . .72
Foam/Water Sprinkler Systems . . . . . . . . . . . . . . . . . .72
Portable and Mobile Fire Protection Options. . . . . . . . . . . . . . . . . . . . . . . .73
Portable Foam Monitors and Handlines. . . . . . . . . . . . . . . . . .73
Portable Foam Wands . . . . . . . . . .75
Mobile Foam Apparatus. . . . . . . . .75
Determining Foam Concentrate Requirements. . . . . . . . . . . . . . . . . . .75
NFPA Recommended Foam Application Rates. . . . . . . . . . . .76
Determining Foam Requirements. . . . . . . . . . . . . . .76
Firewater Supply and Delivery Systems . . . . . . . . . . . . . . . . . . . . . . .80
Water Supply Sources . . . . . . . . . .82
Fixed Pumping and Delivery Systems . . . . . . . . . . . . . . . . . .83
Portable Pumping and Delivery Systems . . . . . . . . . . . . . . . . . .84
Determining Water Supply Requirements. . . . . . . . . . . . . . . . . . .85
Determining Cooling Water Requirements for Exposures. . . . . . . . . . . . . . . . . .88
Summary. . . . . . . . . . . . . . . . . . . . . . . . .89
References and Suggested Readings. . . . . . . . . . . . . . . . . . . . . . .89

CHAPTER 6 **Tactical Response Guidelines . . . . . 90**

Objectives . . . . . 90
Chapter Outline . . . . . 90
Key Terms . . . . . 90
General Response Guidelines . . . . . 91
- Storage Tank Fire Experience . . . . . 91
- General Tactical Guidelines . . . . . 91

Storage Tank Safety Issues . . . . . 92
- Confined Spaces . . . . . 92
- Hydrogen Sulfide . . . . . 94

Boilover, Slopover, and Frothover . . . . . 95
- Boilover . . . . . 95
- Slopover . . . . . 98
- Frothover . . . . . 98

Ground and Dike Fires . . . . . 98
- Size-Up Considerations . . . . . 98
- General Methods of Control and Extinguishment . . . . . 98
- General Strategy and Tactical Options . . . . . 98

Full-Surface Tank Fires . . . . . 100
- Size-Up Considerations . . . . . 100
- General Methods of Extinguishment . . . . . 100
- General Strategy and Tactical Options . . . . . 101

Cone Roof Tank Fires . . . . . 102
- Summary of Construction Features . . . . . 102
- General Methods of Extinguishment . . . . . 102
- Size-Up Considerations . . . . . 102
- General Strategy and Tactical Options . . . . . 103
- Size-Up Considerations . . . . . 103
- General Strategy and Tactical Options . . . . . 104

Open Top Floating Roof Tank Fires . . . . 105
- Summary of Construction Features . . . . . 105
- General Methods of Extinguishment . . . . . 105
- Size-Up Considerations . . . . . 105
- General Strategy and Tactical Options . . . . . 106
- Size-Up Considerations . . . . . 108
- General Strategy and Tactical Options . . . . . 109

Covered Floating Roof Tank Fires . . . . . 109
- Summary of Construction Features . . . . . 109
- General Methods of Extinguishment . . . . . 109
- Size-Up Considerations . . . . . 110
- General Strategy and Tactical Options . . . . . 110

Horizontal and Vertical Low Pressure Tank Fires . . . . . 110
- Summary of Construction Features . . . . . 110
- Size-Up Considerations . . . . . 111
- General Strategy and Tactical Options . . . . . 111
- Size-Up Considerations . . . . . 111
- General Strategy and Tactical Options . . . . . 112

Summary . . . . . 112
References and Suggested Readings . . . 113

Glossary . . . . . 114

Courtesy of William T. Hand.

# Acknowledgments

Over the last 20 years, industrial firefighters have all learned a great deal about how to move large quantities of foam concentrate and water needed for bulk storage tank and refinery firefighting. Back in 1997, much of the current equipment and technology used for both large volume water movement and foam application was not yet developed. Through the innovative efforts of refinery fire chiefs, industrial fire mutual aid groups, private industry, and many others, the business of fighting bulk flammable liquid firefighting has become safer and more effective. Despite these technological improvements, we should not forget that attacking and extinguishing bulk flammable liquids fires is still dangerous work.

The field of hazardous material emergency response becomes more complex and technically detailed every year. A good incident commander has an extensive list of technical specialists in his or her contacts list. To make sure we had the most recent information for our readers, we relied on several experienced industrial firefighters as technical reviewers. These include (alphabetically):

Robert Benedetti, PE, CSP, Flammable Liquids Specialist, National Fire Protection Association, Quincy, MA

David Dean, Assistant Chief of Operations, Refinery Terminal Fire Company, Corpus Christi, TX

Montrell Haldeman, Emergency Response Leader, Monroe Energy, Trainer, PA

Pete Herpst, Refinery Fire Chief (Retired), Flint Hills Resources, Pine Bend, MN

Callen Hight, Assistant Fire Chief, Refinery Terminal Fire Company, Corpus Christi, TX

Roy Rivera, Deputy Fire Chief, Refinery Terminal Fire Company, Port Arthur, TX

Pat Robinson, Fire Chief, PBF Paulsboro Energy, Paulsboro, NJ

The authors would like to give special credit to George Elberti, Safety Director EHS for the Gulf Oil, LP (South Central, PA), for his assistance in acquiring facility and storage tank photographs. His commitment to emergency preparedness and support of the emergency response community is both recognized and sincerely appreciated.

Special thanks to the Refinery Terminal Fire Company Fire Chief Paul Swetish and the RTFC staff who provided various photos.

Finally, throughout the book several illustrations were rendered of fixed fire suppression systems that were based upon illustrations in National Foam's Fire Protection Training Manual. We appreciate the courtesy of allowing us to develop our illustrations based on their work.

# About The Authors



MICHAEL HILDEBRAND and GREGORY NOLL have more than 40 years of emergency preparedness and response experience. During their careers, they have served as firefighters, hazardous materials technicians, instructors, and incident commanders. Both are United States Air Force veteran firefighters and currently serve on a national incident management team.

BILL HAND is a United States Army veteran and is retired from the Houston Fire Department after 31 years of service with the Hazardous Materials Response Team. During his career, he responded to over 7,500 hazmat incidents. Bill has served on numerous technical committees and reviewed and developed hazmat training programs for over 40 years.

All three authors are recipients of the International Association of Fire Chiefs (IAFC) John M. Eversole Lifetime Achievement Award for their leadership and contributions to the hazardous materials emergency response community.

# How to Use This Book



## About the Book

The second edition of *Aboveground Bulk Storage Tank Emergencies* is designed to meet the general requirements for handling aboveground storage tank emergencies as described in the National Fire Protection Association (NFPA) Standard 472—Competence of Responders to Hazardous Materials/Weapons of Mass Destruction Incidents (2018). In addition, the textbook provides the knowledge competencies in Chapter 16—Competencies for Hazardous Materials Technicians with a Flammable Liquids Bulk Storage Specialty.

It should be emphasized that reading this publication will not provide the reader with all of the necessary knowledge to respond to a hazardous materials incident involving bulk flammable liquid storage tanks. For more information consult the textbook, *Hazardous Materials: Managing the Incident*, 4th edition (2014), by Jones & Bartlett Learning.

**The book is divided into six chapters.**

Chapter 1, "Introduction and Overview of Aboveground Storage Tanks," reviews the objectives of the book and examines the storage tank problem.

Chapter 2, "Standards, Codes, and Regulations," takes an in-depth look at the various consensus standards, state and federal regulations which serve as the basis for storage tank design, construction, and operation.

Chapter 3, "Tank Design and Construction Features," provides an overview of the major construction features of each type of storage tank.

Chapter 4, "Incident Management and Response Considerations," reviews the incident management system as it applies to storage tank fires, and provides an overview of the Eight Step Process© as it relates to storage tank emergencies.

Chapter 5, "Firefighting Foam, Water Supply, and Fire Protection Requirements," reviews the basic requirements for selecting an extinguishing agent and establishing a water supply to protect exposures and carry out fire attack and extinguishment.

Chapter 6, "Tactical Response Guidelines," provides some general guidance on how to size up specific types of storage tank fires, evaluate their hazards and risks, and handle various types of tank emergencies.

## Using the Book

This book may be used on an individualized self-study basis, as part of a formal training program, or at emergency services academies.

There are several features of this book you should be familiar with:

**Objectives**: Each chapter provides objectives; those that meet the general requirements of NFPA Standard 472 are indicated.

**Key Terms**: The key terms are defined at the beginning of each chapter when they first appear. A more detailed list of terms is included in the glossary at the end of the book.

**Scan Sheets**: This book contains a lot of information. Some of it is "need to know"; some of it is just "nice to know" or details you might find handy if you are working on a special problem. Scan Sheets are included throughout the book in boxed and screened formats to separate the nice-to-know information from the information you should really know. You can read the entire book without studying the Scan Sheets and still get the highlights.

**Case Studies**: Individual, in-depth Case Studies based on real incidents of historical significance are inserted throughout the book. They are included to help the reader understand the origin of a provision in a code or to reinforce safe operating practices. Like Scan Sheets, you can skip the case studies and still get the highlights by reading the main material.

**Summary**: A brief statement is located at the end of each chapter that summarizes the key points and intent of that unit.

**References and Suggested Readings**: References and suggested readings are included at the end of each chapter.

# Introduction and Overview of Aboveground Storage Tanks

CHAPTER 1

Courtesy of Gregory G. Noll.

## Chapter Outline

- Key Terms
- Introduction
- Target Audience
- Overview of the Storage Tank Industry
- Objectives
- Summary
- Reference

## Key Terms

Aboveground Bulk Storage Tank A horizontal or vertical tank that is listed and intended for fixed installation, without backfill, above or below grade, and is used within the scope of its approval or listing [NFPA 30A].

Barrel A unit of measurement equal to 42 U.S. standard gallons.

Flammable Liquid A liquid whose flash point does not exceed 100°F (37.7 C) when tested by closed-cup test methods. A combustible liquid, on the other hand, is one whose flash point is 100°F (37.7 C) or higher, also when tested. (For further information on classification of liquids and determination of flash point, including appropriate flash point test procedures, see Subsection 3.3.33 and Chapter 4 of NFPA 30.)

Polar Solvent A liquid whose molecules possess a permanent electric moment. Examples are amines, ethers, alcohols, esters, aldehydes, and ketones. In firefighting, any flammable liquid which destroys regular foam is generally referred to as a polar solvent (or is water miscible).

## Introduction

Aboveground bulk storage tanks can be found almost anywhere but are most commonly located at petroleum refineries, bulk storage terminals, pipeline stations, utility plants, and various manufacturing facilities. In addition to the bulk storage tank, these facilities may also include liquid transmission pipelines; piping systems; transfer pumps; additive tanks; and rail, marine, and truck loading racks. When a storage tank spill or fire occurs, it usually involves some or all of this auxiliary equipment (see FIGURE 1-1).

Storage tank fires are rare; while considered a low probability event, they can result in high consequences including loss of life, injury, disruption of the supply chain, significant financial loss of inventory and equipment, and environmental impacts. Storage tank fires typically require big foam concentrate and water resources and an incident management team to bring the problem under control. These fires are impressive and they typically generate lots of local or national media coverage.

Storage tank emergencies can involve a wide range of products and tank designs in many different configurations. Despite the wide variety of storage tank types and

FIGURE 1-1 Aboveground storage tanks can be found almost anywhere in a variety of types and built to national standards. Construction of new and larger storage tanks is occurring along the U.S. energy corridors.

Courtesy Gulf Oil Limited Partnership

possible emergency scenarios, the majority of storage tank emergencies involve flammable liquids. Therefore, we have limited the scope of this book to storage tank emergencies involving both Class 3 hydrocarbons and polar solvents (see FIGURE 1-2).

This textbook will primarily focus on the five basic types of bulk petroleum storage tanks including the cone (fixed) roof tank, open (external) floating roof tank, covered (internal) floating roof tank, vertical low pressure storage tank, and horizontal low pressure storage tank. The scope of this book does not include underground storage tanks or caverns, nor does it address bulk flammable gas storage tank emergencies.

FIGURE 1-2 Aboveground storage tanks can hold numerous hazard classes, but most emergencies involving tanks hold a Class 3 liquid such as crude oil, gasoline, or ethanol.

Courtesy of William T. Hand.

## Target Audience

The target audience for this textbook is emergency responders who are responsible for planning and/or responding to bulk flammable liquid storage tank emergencies. These may include public safety and industrial firefighters, hazardous materials technicians, and incident commanders (see FIGURE 1-3).

In addition, technical specialists who may provide strategic and tactical-level recommendations to the incident commander and other responders may find the textbook of value.

## Overview of the Storage Tank Industry

The aboveground bulk storage tank is the workhorse of the energy industry. Such tanks are found in all phases of the petroleum production, refining, transportation, and distribution network. From storing crude oil to finished refined products, these tanks are an integral element of the North American economy.

The actual number of aboveground bulk storage tanks is unknown; however, the U.S. Energy Information Administration measures crude oil storage capacity twice each year. From September 2015 to March 2016, the United States added 34 million barrels (6%) of working crude oil capacity. More storage tank capacity is being added across the United States, especially in the Gulf Coast and Midwest states. As of January 2016, there was a crude commercial

FIGURE 1-3 Storage tank facilities should be visited and preplanned by the first due fire companies to ensure familiarization with the fire protection features and requirements.

Courtesy of Gregory G. Noll.

storage volume of 532 million barrels across the United States! All of that crude oil is maintained in aboveground bulk storage tanks. This inventory does not include the 695 million barrels the U.S. Department of Energy maintains in the U.S. Strategic Petroleum Reserve, which is the largest reserve in the world.

Despite the large storage tank population and their exposure to a variety of harsh operating conditions, the safety record of the aboveground bulk storage tank has been very good (see **FIGURE 1-4**).

Over the decades, much has been learned about tank fabrication, welding, construction and inspection practices, design of pressure relief devices, among other things. The current storage tank safety standards are based on years of operating experience and have incorporated lessons learned from past accidents into reliable engineering standards, safe operating practices, and training programs.

Some general observations that can be made concerning aboveground storage tanks are as follows:

- Petroleum storage tanks are not just found in the petroleum industry. The fact is that aboveground storage tanks are widely owned and operated by schools, public utilities, municipal bus fleets, construction companies, farm cooperatives, and the military, among many other businesses. If you focus on preincident planning at only petroleum industry facilities and do not look at the bigger picture, you will miss a lot of tanks and potential risks in your community.
- The petroleum storage tank owner and operator safety record in preventing spills, fires, and catastrophic tank failures has been very good and is often cited in industry and government studies and reports. When you consider the millions of tanks in service and the potential exposure they represent to the community, storage tanks typically pose a lower risk when compared to the scenarios created by other types of both natural and technological hazards found in the community.

Despite the good safety record of owners and operators in *preventing* fires and spills, historically petroleum storage tank incidents have killed and injured many firefighters. Over 70 firefighters have been killed in the line of duty in the last 50 years at incidents involving flammable liquid storage tank fires. Some of these incidents involved multiple fatalities at a single incident. Fortunately, the current risk-based response processes used by the emergency response community have vastly improved firefighter safety. As of 2017, we are unaware of any firefighter fatalities from a storage tank emergency in the last 15 years. But history can repeat itself if we forget the lessons learned (see **FIGURE 1-5**).

Aboveground bulk storage tank fires are complex incidents that require a lot of resources to bring the incident under control. When one takes an objective look at the history of fighting large flammable liquid fires, the costs can be very high. Some of the largest losses in the history of the hydrocarbon and chemical industries have involved petroleum storage tanks. See Scan Sheet 1-A for several examples.

**FIGURE 1-4** Storagetanks are designed to very high engineering standards and have a good safety record despite the harsh operating environments where they are located.

Courtesy of William T. Hand.

**FIGURE 1-5** Over 70 firefighters have been killed in the line of duty in the last 50 years at incidents involving flammable liquid storage tank fires. Some of these incidents involved multiple fatalities at a single incident.

Refinery Terminal Fire Company

# Scan Sheet 1-A—Summary of the Three Largest Petroleum Storage Tank Losses in the United States

Note: Original loss figures are shown by date. Numbers in parentheses have been converted to 2017 dollars to adjust for inflation. This helps the reader understand how expensive these losses would be in current dollars.

**November 25, 1990**
**Denver, Colorado**
**$32,000,000 ($59,000,000) Loss**

A fire at a 16-acre tank farm that supplied jet fuel to the Denver International Airport burned for more than 55 hours, damaging seven storage tanks and consuming more than 1.6 million gallons of jet fuel. The tank farm was comprised of 12 storage tanks.

At approximately 9:20 a.m., the fuel supply company received a "no flow" indication in the pipeline to the tank farm. Shortly after that time, the airport control tower noticed a black column of smoke from the tank farm. An initial fuel leak originating at an operating fuel pump in the valve pit was ignited by the electric motor for the pump resulting in the fire. A cracked fuel supply pipe in the valve pit formed two V-shaped streams extending 25 to 30 feet into the air. This provided additional fuel to the pool fire.

As the fire continued to grow, coupling gaskets in the piping deteriorated and more fuel flowed out of the storage tanks, spreading the fire throughout the dike area. The valve controlling the fuel flow to the airport supply line sporadically released fuel into the valve pit. Firefighters were unable to prevent the back flow of fuel from this line since the nearest manual shutoff valve was 2 miles from the tank farm.

The Denver Airport Fire Department dispatched four aircraft rescue firefighting (ARFF) vehicles and one rapid intervention vehicle to the fire. The second and third alarms provided an additional five engine companies, three truck companies, and one rescue unit from the Denver City Fire Department.

In addition to the foam concentrate on hand at the scene, more was received from other local departments as well as the Seattle, Houston, and Chicago fire departments. Unknown to the fire department, a pipeline that was reported to have been shut down continued to supply fuel to the fire. After repeated unsuccessful attempts to extinguish the fire by the Denver Fire Department, Williams Fire and Hazard Control was brought in by the owners to assist the fire department with extinguishing the fire. The pipeline was eventually shut down and the fire was extinguished.

**July 7, 1983**
**Newark, New Jersey**
**$35,000,000 ($85,000,000) Loss**

Gasoline was being received by pipeline into a 42,000-barrel internal floating roof tank at a products terminal when an overfill occurred, spilling about 1,300 barrels into the tank diked area. A slight wind (1 to 5 mph) carried gasoline vapors about 1,000 feet to a drum reconditioning plant, where an incinerator provided the ignition source. The resulting explosion caused $10,000,000 damage to the terminal and up to $25,000,000 in over 2,000 damage claims to rail cars and adjacent properties. Although dikes contained the burning spill to the tank that was overfilled, two adjoining internal floating roof tanks and a smaller tank ignited and were eventually destroyed along with 120,000 barrels of product. Since the burning tanks presented little exposure to other facilities, the decision was made to let the fire burn itself out. This incident resulted in tank overfill prevention requirements in NFPA-30, The Flammable and Combustible Liquids Code.

**September 24, 1977**
**Romeoville, Illinois**
**$8,000,000 Loss ($32,000,000)**

Lightning struck a 190-foot-diameter cone roof tank containing diesel fuel. Roof fragments were thrown 240 feet and struck a 100-foot-diameter covered floating roof gasoline tank. An adjacent, 180-foot-diameter floating roof gasoline tank 80 feet away was also struck by debris. The entire surfaces of the cone and internal floating roof tanks ignited immediately. The rim fire on the floating roof resulted in the roof sinking after about 4 hours. The two largest tanks were full; the smallest about half full. The two larger tanks and their contents were destroyed. The fire in the internal floating roof tank was extinguished after about 2 hours.

After burning for approximately 46 hours, the fire was extinguished through both top-side and subsurface foam applications. The refinery's five stationary fire pumps supplied up to 10,000 gpm of the estimated 14,000 gpm required during the fire. Thirty-five municipal and industrial fire departments, including a 12,000-gpm fire boat, assisted the refinery fire department. About 22,000 gallons of foam concentrate were consumed during the fire-fighting effort.

Sources:
Adapted from Marsh, *The 100 Largest Losses 1974–2015: Large Property Damage Losses in the Hydrocarbon-Chemical Industries* (24th edition).

## Summary

Aboveground bulk storage tanks can be found almost anywhere but are most commonly located at petroleum refineries, bulk storage terminals, pipeline stations, utilities plants, and various manufacturing facilities. In addition, aboveground storage tanks are widely owned and operated by schools, public utilities, municipal bus fleets, construction companies, farm cooperatives, the military, among many other businesses.

Storage tanks have a good safety record, and major fires are rare. They are considered low probability events but can result in high consequences including loss of life, injury, disruption of the supply chain, significant financial loss of inventory and equipment, and environmental impacts.

This book will address five basic categories of storage tanks including the cone (fixed) roof tank, open (external) floating roof tank, covered (internal) floating roof tank, vertical low pressure storage tank, and horizontal low pressure storage tank.

The target audience for this textbook is emergency responders who are responsible for planning and/or responding to bulk flammable liquid storage tank emergencies. These may include public safety and industrial firefighters, hazardous materials technicians, and incident commanders.

In addition, technical specialists who may provide strategic and tactical-level recommendations to the incident commander and other responders may find the textbook of value.

## References and Suggested Readings

1. National Fire Protection Association, *NFPA 30—Flammable and Combustible Liquids Code* (2016 edition). Quincy, MA: National Fire Protection Association (2016).

CHAPTER

2

# Standards, Codes, and Regulations

Courtesy of William T. Hand.

## Chapter Outline

- Objectives
- Key Terms
- Introduction
- Consensus Standards and Codes
- National Fire Protection Association
- American Petroleum Institute
- Underwriters Laboratories, Inc.
- Steel Tank Institute
- Regulations
- Summary
- References

## Objectives

1. List and describe the primary consensus standards, codes, and regulations that address the design, construction, and operation of aboveground liquid petroleum storage tanks.

## Key Terms

American Petroleum Institute (API) National trade association that represents all aspects of America's oil and natural gas industry.

American Society of Mechanical Engineers (ASME) A not-for-profit professional organization that enables collaboration, knowledge sharing, and skill development across all engineering disciplines, while promoting the vital role of the engineer in society.

National Fire Protection Association (NFPA) An international voluntary membership organization to promote improved fire protection and prevention, establish safeguards against loss of life and property by fire, and write and publish national voluntary consensus standards.

Steel Tank Institute (STI) A not-for-profit organization that works with tank manufacturers, users, regulatory authorities, and consultants to promulgate standards for the design, construction, and installation of aboveground and underground tanks used for the storage of flammable and combustible liquids. The STI publishes consensus standards that guide tank manufacturers and code enforcement officials.

Underwriters Laboratories (UL) An organization that helps companies demonstrate safety, confirm compliance, enhance sustainability, manage transparency, deliver quality and performance, strengthen security, protect brand reputation, build workplace excellence, and advance societal well-being. Some of the services offered by UL include inspection, advisory services, education and training, testing, auditing and analytics, certification software solutions, and marketing claim verification.

## Introduction

Standards and regulations provide guidance to manufacturers, owners and operators, and code enforcement officials on the safe design, operation, inspection, and maintenance of aboveground flammable and combustible liquids storage tanks (see FIGURE 2-1).

This chapter will provide an overview of the primary voluntary consensus standards, codes, and regulations that govern all types of aboveground storage tanks.

## Consensus Standards and Codes

Voluntary consensus standards are developed through professional organizations or trade associations as a method of improving the individual quality of a product or system. In the United States, standards are developed through

FIGURE 2-1 Aboveground storage tanks are governed by numerous standards and regulations. Many safety-related consensus standards are also adopted by reference in a regulation, such as when a federal, state, or municipal government adopts a consensus standard by reference.

Courtesy of William T. Hand.

a consensus process whereby a committee of technical specialists representing varied interests writes the first draft of the standard. The document is then submitted to either a larger body of specialists or the general public who then may amend, vote on, and approve the standard for publication. This procedure is known as the Voluntary Consensus Standards Process.

When a consensus standard is completed, it may be voluntarily adopted by government agencies, individual companies and organizations, or industry-specific professional trade associations. Many safety-related consensus standards are also adopted by reference in a regulation, such as when a federal, state, or municipal government adopts a consensus standard by reference.

Section 12 of the National Technology Transfer and Achievement Act of 1996 [15 U.S.C. §3701 et seq. (1996 and Revised 2016)] states: ". . . all federal agencies and departments shall use technical standards that are adopted by voluntary consensus standards as a means to carry out policy objectives or activities determined by the agencies or departments." Under the act, technical standards are defined as "performance-based or design-specific technical specifications and related management system practices."

Standards developed through the voluntary consensus process play an important role in increasing both workplace and public safety. Historically, a voluntary standard improves over time as each revision reflects recent field experience and adds requirements that reflect changes in technology and risks. As users of the standard adopt it as a way of doing business, the level of safety gradually improves.

A voluntary consensus standard provides a way for individual organizations and corporations to self-regulate their business or profession. All of the national fire codes in the United States are developed through the voluntary consensus standards process, with the National Fire Protection Association (NFPA) being a key player.

## National Fire Protection Association

The National Fire Protection Association (NFPA) is an international voluntary membership organization to promote improved fire protection and prevention, establish safeguards against loss of life and property by fire, and write and publish national voluntary consensus standards.

NFPA has many individual hazardous materials–related voluntary consensus standards. Several of these deal specifically with flammable and combustible liquids petroleum storage tank issues. These standards are developed through the committee process according to NFPA rules and procedures and are usually revised about every 5 years.

NFPA's hazardous materials standards are widely used by both industry and public safety organizations as recommended practices for inspection, safe handling and installation, among other things. They do not have the force of law unless they are adopted by a government agency that has power of enforcement. For example, state governments which have approved state OSHA plans under section (18b) of the Occupational Safety and Health Act of 1970 must adopt standards to enforce requirements that are at least as effective as federal requirements. Many U.S. states have adopted federal OSHA regulations as state law. While NFPA does not specifically write regulations, its standards often end up as law through an adoption process at the federal and state levels.

There are two NFPA standards that are widely used in the code enforcement area as it pertains to flammable and combustible liquids: NFPA 1 and NFPA 30.

**NFPA 1: Fire Code** provides a comprehensive, integrated approach to fire code regulation and hazard management. The code features extracts and references from more than 130 NFPA codes and standards, such as the NFPA 101: Life Safety Code; NFPA 54: National Fuel Gas Code; NFPA 58: Liquefied Petroleum Gas Code; NFPA 25: Inspection, Testing, and Maintenance of Water-Based Fire Protection Systems; NFPA 13: Installation of Sprinkler Systems; NFPA 72: National Fire Alarm and Signaling Code; and NFPA 30: Flammable and Combustible Liquids Code (see FIGURE 2-2).

**NFPA 30: Flammable and Combustible Liquids Code** is the primary fire safety code for the storage and

**FIGURE 2-2** NFPA1: Fire Code and NFPA 30: Flammable and Combustible Liquids Code are both used to guide fire protection engineers and fire protection inspection teams. Both standards are frequently referenced in inspection and audit reports to ensure a high level of reliability and operational readiness. Many regulatory agencies adopt the standards as law.

Courtesy of Michael S. Hildebrand.

handling of flammable and combustible liquids. Of the code's chapters, six govern the bulk storage of flammable and combustible liquids in both tank and piping systems. One chapter governs tank railcar and tank truck loading/unloading facilities. The code also addresses container storage in rooms and warehouses, and the handling of liquids in situations ranging from incidental uses to large chemical plants and petroleum refineries.

The NFPA 30 chapters on storage tanks and piping systems include the following topics:

- Design and construction of aboveground and underground tanks
- Requirements for siting and spacing tanks from each other and from adjacent buildings, property lines, and public roads
- Spill control including diking and remote impounding and combinations thereof, closed-top diking, and secondary containment-type tanks
- Sizing of normal breather vents and of emergency relief vents
- Tank supports and anchoring
- Testing and maintenance
- Overfill protection
- Piping requirements and arrangements

Additional information on NFPA standards, recommended practices, and related documents can be found at www.nfpa.org.

## American Petroleum Institute

The American Petroleum Institute (API) is a national trade association that represents all aspects of America's oil and natural gas industry. Corporate members include the largest integrated major oil companies to the smallest of independents and come from all segments of the industry. They are producers, refiners, suppliers, pipeline operators, and marine transporters, as well as service and supply companies that support all segments of the industry. API has several aboveground storage tank standards which are summarized in Scan Sheet 2-A.

# Scan Sheet 2-A—American Petroleum Institute Storage Tank–Related Safety Standards

- **API Standard 12B: Specifications for Bolted Tanks for Storage of Production Liquids**
  Covers material, design, and erection requirements for vertical cylindrical aboveground bolted steel production tanks in nominal capacities of 100 barrels to 10,000 barrels (in standard sizes). It also contains appurtenance requirements.
- **API Standard 12D: Specification for Field Welded Tanks for Storage of Production Liquids**
  Covers material, design, fabrication, and erection requirements for vertical cylindrical aboveground welded steel production tanks in nominal capacities of 500 barrels to 10,000 barrels (in standard sizes).
- **API Standard 12F: Specification for Shop Welded Tanks for Storage of Production Liquids**
  Covers materials, design, and construction requirements for vertical cylindrical aboveground shop welded steel production tanks in nominal capacities of 90 barrels to 500 barrels (in standard sizes up to a maximum diameter of 16 feet).
- **API Standard 620: Design and Construction of Large, Welded, Low Pressure Storage Tanks**
  Covers the design and construction of large, welded, field-assembled storage tanks used to store petroleum intermediates and finished products operated at a gas pressure of 15 pounds per square inch gauge and less.
- **API Standard 2000: Venting Atmospheric and Low Pressure Storage Tanks**
  Covers recommended procedure for testing venting devices on low pressure aboveground storage tanks used for petroleum and petroleum products. It also contains venting-capacity tables.
- **API Recommended Practice 2001: Fire Protection in Refineries**
  Provides a better understanding of fire protection problems and of the steps necessary to ensure the safe storing, handling, processing, and shipping of petroleum and petroleum products in refineries. The general principles mentioned in the publication are based in large measure upon composite experience in a large number of refineries.
- **API Recommended Practice 2003: Protection Against Ignitions Arising Out of Static, Lightning, and Stray Currents**
  Covers some of the conditions that have resulted in oil fires ignited by electrical sparks and arcs from natural causes, as well as the methods the petroleum industry currently is applying for the prevention of ignition from these sources.
- **API Publication 2009: Safe Welding, Cutting, and Hot Work Practices in Gas and Electrical Cutting and Welding in the Petroleum and Petrochemical Industries**
  Outlines suggested precautions for the protection of persons from injury and for the protection of property from injury and for the protection of property from damage by fire which might arise during the operation of gas and electrical cutting and welding equipment in and around petroleum operations.
- **API Standard 2015: Requirements for Safe Entry and Cleaning of Petroleum Storage Tanks**
  Covers discussion of safe practices entering and tank cleaning of storage tanks, including use of suitable mechanical equipment and protective clothing, use of proper cleaning methods, elimination of potential ignition hazards, and provision for means of entrance and exit in an emergency.
- **API Recommended Practice 2021: Management of Atmospheric Storage Tanks**
  Is designed as a guide to train employees to successfully attack and fight petroleum tank fires.

- **API Recommended Practice 2023: Guide for Safe Storage and Handling of Heated Petroleum-Derived Asphalt Products and Crude Oil Residua**
  Serves as a guide to precautions to be followed by personnel for the storage and handling of asphalt products when stored in heated tanks.
- **API Publication 2517: Evaporative Loss from External Floating Roof Tanks**
  Presents a method for estimating evaporation losses from floating-roof tanks containing multicomponent hydrocarbon mixtures.
- **API Standard 2550 (ASTM D 1220-65): Measurement and Calibration of Uprighted Cylindrical Tanks**
  Covers procedures for calibrating upright cylindrical tanks larger than a barrel or drum, including procedures for making necessary measurements to determine total and incremental tank volumes and the recommended procedure for computing volumes.
- **API Standard 2555 (ASTM D 1406-65): Method for Liquid Calibration of Tanks**
  Covers standard procedure for calibrating tanks, or portions of tanks, larger than a barrel or drum by introducing or withdrawing measured quantities of liquid.
- **Guide for Inspection of Refinery Equipment: Chapter XIII, Atmospheric and Low Pressure Storage Tanks**
  Covers inspection of atmospheric storage tanks which have been designed for operation at pressures from atmospheric through 0.5 pound per square inch gauge and inspection of low pressure storage tanks which have been designed for operation at pressures above 0.5 pound per square inch gauge but not exceeding 15 pounds per square inch gauge.

  To learn more about American Petroleum Institute Publications visit api.org.

## API-650: Welded Steel Storage Tanks for Oil Storage

API-650 is the most widely recognized standard for petroleum storage tanks in the world. The standard has been adopted by reference in NFPA 30 and many state fire codes.

API-650 is based on over 70 years of accumulated operating experience in the petroleum industry and manufacturers of welded steel storage tanks. The primary objective of the standard is to provide a purchase specification to facilitate the manufacture and procurement of storage tanks.

The standard covers design, fabrication, erection, and testing requirements for vertical cylindrical aboveground storage tanks. API-650 covers a wide variety of storage tank types including closed and open top welded steel tanks in various sizes and capacities for internal pressures which are generally near atmospheric. The standard does not cover tanks in refrigerated service.

The API-650 standard provides industry and the public with tanks of adequate safety and reasonable economy. The standard does not establish a fixed tank size or capacity but does provide detailed recommendations for tank specification and design. The standard covers the following major areas:

- **Materials.** Materials used in the construction of tanks must conform to the latest edition of the standard or one of its accepted referenced standards (e.g., ASTM, ASME). API-650 includes detailed requirements that all steel plates for tank shells, roofs, and bottoms meet high-quality standards. All steel used in API-650 tanks must be manufactured by the open hearth, electric furnace, or basic oxygen process. Shell plates are limited to a maximum thickness of 1.75 inches unless a lesser thickness is specified. Plates over 1.5 inches must be normalized or quench tempered, kilned, made to fine grain practice, and impact tested.
- **Design.** API-650 includes a comprehensive list of tank design requirements for use in developing tank purchase and manufacturing specifications. Design criteria include welding specifications, design factors such as external loads, shell and bottom plate specifications, top and wind girder design, roof design, and special requirements for corrosion control.

  The design criteria include a requirement that all roofs and supporting structures must be able to support a dead load, plus a uniform live load of not less than 25 pounds per square foot. Roof plates must have a minimum nominal thickness of 3/16 inch. A greater thickness is required for self-supporting roofs. This is an important safety requirement, because firefighters may need to walk onto the top of a tank carrying firefighting equipment providing that the tank is well maintained and undergoes periodic inspections.

  Another important design requirement in API-650 is what is referred to as a "weak roof-to-shell seam."

The standard includes specific requirements for weld thicknesses which will permit the fixed tank roof to fail by design in the event of an overpressure situation. For example, if an internal explosion occurs from a lightning strike, the tank is designed to fail at the roof-to-shell seam and relieve the pressure at the top of the tank. This design feature helps minimize the violent failures experienced in older tanks which sometimes failed at the shelf-to-bottom plate seam creating a rocket.

- **Fabrication.** Even when a storage tank is manufactured according to design specifications, faulty workmanship in the shop can create serious safety problems which may go undetected to the purchaser or the inspector. All work fabricating API-650 tanks must be done in accordance with the requirements of the standard. The workmanship and finish of API-650 tanks and its components must be first class in quality. API requires that fabrication shops producing an API-650 specification tank must permit the purchasing inspectors free entry to all parts of the manufacturer's shop whenever any work under contract is being performed. The manufacturer is also required to furnish the purchaser samples for offsite quality testing when requested.
- **Erection.** The standard includes detailed requirements for erection of the completed tank onsite. Requirements include welding procedures, inspection, testing, repair procedures, inspection enforcement, and measurements.
- **Methods of Inspecting Joints.** The quality of welds is a critical aspect of storage tank safety. Weld joints are a special concern. API-650 specifies radiographic and ultrasonic inspection techniques for inspecting joints.
- **Welding Procedures and Welding Qualifications.** API-650 requires that storage tank welding procedures meet the requirements of the American Society of Mechanical Engineers (ASME). Each welder working on the project must be assigned an identifying number or symbol by the fabrication or erection manufacturer. A record of the welder's work is required showing the date and results of tests and identifying marks. These records must be certified and be available for inspection.
- **Marking Tanks.** Tanks made in accordance with API-650 must be identified by a nameplate welded to the tank shell adjacent to a manway. The nameplate must indicate that the tank meets the API-650 requirements and contain specific information (see FIGURE 2-3).

**FIGURE 2-3** The specification plate identifies that the tank has been manufactured to the API-650 standard and meets the design and engineering requirements that incorporate various safety features. This plate indicates that the tank was built to the API-650 2008 edition, has a diameter of 104 feet, as well as other information.

Courtesy of William T. Hand.

## API-620: Standard for Design and Construction of Large, Welded, Low Pressure Storage Tanks

API-620 covers the design and construction of large, welded, low pressure carbon steel aboveground storage tanks. Tanks constructed to API-620 standards handle internal pressures not exceeding 15 pounds per square inch gauge and are intended to hold or store liquids with gases or vapors above the surface of the liquid or to store vapors without liquid in the tank.

API-620 is limited to vertical, cylindrically shaped tanks. It does not cover the design and construction of horizontal tanks; however, nothing precludes the application of similar design criteria to horizontal tankage.

The standard covers the following major areas:

- **Materials.** Materials used in the construction of API-620 tanks must meet rigid specifications. All steel materials must be capable of withstanding contact with liquids of 250°F (121°C) with pressures of up to 15 psig (103.43 KPa). Materials must also withstand cold weather temperatures where the lowest recorded one-day mean temperature is −50°F (−45.5°C)
- **Design.** API-620 includes detailed engineering and design requirements and provides the fundamental rules which can be used as a basis for an inspector to judge the safety of any tank containing the API

nameplate. (See the description of the API nameplate under API-650.)

API-620 requires that the volume of the vapor space above the high liquid level upon which the nominal capacity is based must not be less than 2% of the total liquid capacity. The standard also requires that tanks be capable of handling routine internal pressure and vacuum which may be caused by normal temperature fluctuations.

- **Fabrication.** API-620 includes comprehensive requirements for fabrication of storage tanks. These requirements are similar in scope to those outlined under API-650.
- **Inspection and Testing.** After tank erection has been completed, API-620 requires that the tank pass a series of hydrostatic and pneumatic tests. API requires that all attachment welding around all openings must be inspected by the magnetic-particle method, both inside and outside. When the underside of a tank is not accessible after erection, this phase of the inspection may be omitted.

  Following magnetic-particle inspection, air pressure at 15 psig (103.43 KPa) must be introduced between the tank wall and the reinforcing plate, saddle flange, or integral reinforcing pad. While each space is subjected to pressure, a solution film must be applied to all attachment welding around the reinforcement, both inside and outside.

  Tanks that have been designed and constructed to be filled with liquid to the top of the roof must be subjected to a hydrostatic test. Basically, the tank is filled with water and allowed to set for 1 hour.
- **Pressure and Vacuum Relieving Devices.** API-620 tanks must be protected by automatic pressure-relieving devices that will prevent the pressure at the top of the tank from rising more than 10% above the maximum allowable working pressure.

  Whenever the tank can be subjected to direct flame exposure, supplemental pressure-relieving devices must be installed. These devices must be capable of preventing the pressure from rising more than 20% above the maximum allowable working pressure (MAWP). A single pressure-relieving valve may be used if it meets special requirements outlined in the standard.

  Vacuum relieving devices must also be installed to permit the entry of air or other gases and vapors to avoid collapse of the tank wall. These devices must be located above the liquid level in the tank so that they can function properly.

## Underwriters Laboratories Inc.®

Underwriters Laboratories (UL) is a not-for-profit organization formed in 1894 whose mission is to establish, maintain, and operate laboratories for the examination and testing of devices, systems, and materials to determine their relation to hazards to life and property; and to ascertain, define, and publish standards, classifications, and specifications for materials, devices, products, equipment, construction, methods, and systems affecting such hazards.

UL standards for safety are developed under a procedure that provides for participation and comment from the affected public as well as industry. The procedure takes into account a survey of known existing standards and the needs and opinions of a wide variety of interests concerned with the subject matter of the standard (e.g., flammable and combustible liquid storage equipment). Thus, manufacturers, consumers, individuals associated with consumer groups, government officials, industry representatives, inspection authorities, insurance interests, and others provide input to UL in formulating safety standards to keep them in step with technological advances.

The flammable and combustible liquid storage tank standards UL published are summarized in Scan Sheet 2-B.

## Steel Tank Institute

The Steel Tank Institute (STI) is a not-for-profit organization formed in 1916 to secure cooperative action in advancing the common purposes of its members to promote activities designed to enable the industry to conduct itself with the greatest economy and efficiency. STI cooperates with other industries, organizations, and government bodies in the development of reliable standards, which advance industry manufacturing techniques to solve market-related problems.

The STI works with tank manufacturers, users, regulatory authorities, and consultants to promulgate standards for the design, construction, and installation of aboveground and underground tanks used for the storage of flammable and combustible liquids. See Scan Sheet 2-C.

# Scan Sheet 2-B—Underwriters Laboratories–Related Storage Tank Safety Standards

- **UL 80: Steel Inside Tanks for Oil Burner Fuels and Other Combustible Liquids**

  Covers steel diked–type atmospheric storage tanks from 60 to 660 gallons (227 to 2,500 L) intended primarily for the storage and supply of heating fuels for oil burning equipment, or alternately for the storage of diesel fuels for compression ignition engines and motor oils (new and used) for automotive service stations, in aboveground applications.
- **UL 142: Steel Aboveground Flammable Liquid Tanks**

  Covers horizontal and vertical welded steel tanks intended for the outside storage aboveground of flammable and combustible liquids at pressures in vapor spaces between atmospheric and 0.5 psig (3.45 KPa). They are intended for use with only noncorrosive, stable liquids that have a specific gravity not exceeding that of water. Tanks covered by these requirements are cylindrical in shape and are constructed, inspected, and tested for leakage before being shipped from the factory as completely assembled vessels. They are intended for stationary installations in accordance with NFPA 30: Flammable and Combustible Liquids Code. These requirements do not apply to tanks covered by API-650, nor to tanks intended for use in chemical and petrochemical plants.
- **UL 2080: Standard for Fire Resistant Tanks for Flammable and Combustible Liquids**

  Covers shop fabricated, aboveground atmospheric fire-resistant tanks intended for storage of stable flammable or combustible liquids that have a specific gravity not greater than 1.0 and that are compatible with the material and construction of the tank. Tanks meeting this standard are intended to limit the heat transferred to the primary tank when the construction is exposed to a 2-hour hydrocarbon pool fire. Tanks appropriately identified by product markings provide protection for the primary tank against projectile impact and vehicle impact.
- **UL 2085: Standard for Protected Aboveground Tanks for Flammable and Combustible Liquids**

  Covers insulated aboveground atmospheric tanks intended for aboveground storage of noncorrosive, stable, flammable or combustible liquids that have a specific gravity not exceeding that of water.

  These tank construction requirements include an insulation system which is intended to reduce the heat transferred to the primary tank if the construction is exposed to a hydrocarbon pool fire. Tanks covered by these requirements are fabricated, inspected, and tested for leakage before shipment from the factory as completely assembled units. Fire-resistant tanks are intended for installation in accordance with NFPA 30: Flammable and Combustible Liquids Code and NFPA 30-A: Code for Motor Fuel Dispensing Facilities and Repair Garages.

# Scan Sheet 2-C—Related STI Standards

- **F911: Standard for Diked Aboveground Storage Tanks**
  Addresses fabrication specifications for construction of steel dike secondary containment of product(s), with a specific gravity of 1.0 or less, stored in aboveground storage tanks. It includes the manufacture, inspection, and testing of such aboveground secondary containment systems prior to shipment. The standard does not address pipe connections to tank fittings.
- **F941: Fireguard™ Thermally Insulated Aboveground Storage Tank Standard**
  Covers a method of thermally insulating an aboveground storage tank for the purpose of providing a 2-hour fire rating. This design utilizes a double-wall steel tank with a monolithic insulation placed into the interstice of the tank. Both the primary and the secondary tank are equipped with emergency vents. The tank assembly is fully tested with Underwriters Laboratories in accordance with UL 2085 and has a listing for an insulated secondary containment aboveground tank for flammable liquids. The interstitial space has been tested to ensure that a fluid will flow and be detectable and the emergency venting will work properly in the event of a pool fire.
- **F921: Standard for Aboveground Tanks with Integral Secondary Containment**
  Addresses aboveground steel secondary Type I containment (atmospheric type) vessels with built-in monitoring capability for the purpose of giving advance notice to avid environmental contamination.
  The primary intent of the standard is to address the petroleum product storage segment, although storage of other noncorrosive, stable liquid chemicals may be covered. It includes the manufacture, inspection, and testing of such secondary containment tanks prior to shipment.

## Regulations

### U.S. Environmental Protection Agency (EPA)

#### Clean Water Act

The EPA has the authority to regulate aboveground storage tanks under the Clean Water Act. Specifically, the act authorizes the EPA to issue regulations establishing procedures, methods, equipment, and other requirements to prevent the discharge of oil into navigable waterways.

The two most significant storage tank regulations include the Spill Prevention, Control, and Countermeasures (SPCC) requirements under 40 CFR 112, and the Oil Pollution Act (OPA) of 1990.

#### Spill Prevention, Control, and Countermeasure Plan

This regulation was issued by the EPA in 40 CFR 112 and applies to facilities engaged in drilling, producing, gathering, storing, processing, refining, transferring, distributing, or consuming oil and oil products, which due to location could reasonably be expected to discharge oil in quantities that may be harmful into or upon navigable waterways or adjoining shorelines, or upon the waters on the contiguous zone.

Owners or operators of these facilities must prepare an SPCC Plan that includes appropriate containment and/or drainage control structures or equipment to prevent discharged oil from reaching navigable water before cleanup occurs.

Facilities that are excluded from being required to have an SPCC Plan are facilities that do not have aboveground storage capacity of at least 1,320 gallons of oil. No single container at these facilities can hold more than 660 gallons of oil.

#### Oil Pollution Act of 1990 (OPA)

Commonly referred to as OPA-90, this law amended the Federal Water Pollution Control Act. Its scope covers both facilities and carriers of oil and related liquid products, including deepwater marine terminals, marine vessels, pipelines, and railcars. Requirements include the development of emergency response plans, regular training and exercise sessions, and verification of spill resources and contractor capabilities. The law also requires the establishment of Area Committees and the development of Area Contingency Plans (ACPs) to address oil and hazardous substance spill response in coastal zone areas.

### U.S. Occupational Safety and Health Administration (OSHA)

OSHA is the lead federal agency responsible for protecting the safety and health of workers in the workplace. This statutory authority is derived from the Occupational Safety and Health Act of 1970. This law grants OSHA the authority to develop safety and health regulations and to conduct inspections to identify unsafe work practices. It requires that corrective action be taken to protect worker health and safety. The law also allows OSHA to issue citations and pursue civil and criminal prosecution for noncompliance with its regulations.

FIGURE 2-4 OSHA's flammable and combustible liquids regulations are based on the NFPA Flammable and Combustible Liquid code, which requires inspection and maintenance of fire suppression equipment and systems.

Courtesy of Michael S. Hildebrand.

OSHA regulations cover a wide variety of industries, processes, and environments, including facilities that manufacture, use, and store flammable and combustible liquids. OSHA's flammable and combustible liquid regulations are found in 29 CFR 1910.106 (see FIGURE 2-4).

#### 29 CFR 1910.106: Flammable and Combustible Liquids

OSHA's requirements for flammable and combustible liquid storage and use are based on NFPA 30: Flammable and Combustible Liquids Code. These requirements include a number of design and construction requirements for aboveground and underground tanks for flammable and combustible liquid service, as well as the proper installation procedures.

The design and construction requirements contained in the OSHA regulations are referenced from both UL and API tank standards. The standard may be adopted by government agencies, individual corporations, or other organizations and become a regulation.

## Summary

Voluntary consensus standards are developed through professional organizations or trade associations as a method of improving the individual quality of a product or system. Standards are developed through a consensus process whereby a committee of technical specialists representing varied interests writes the standard and then submits it to either a larger body of specialists or the general public who then may amend, vote on, and approve the standard for publication.

NFPA-30 is the primary fire safety code for storage and handling of flammable and combustible liquids. The code's five main chapters govern bulk storage of liquids in tank and piping systems. The code also addresses container storage in rooms and warehouses, and the handling of liquids in situations ranging from incidental uses to chemical plants and petroleum refineries.

The American Petroleum Institute (API) is a national trade association that represents all aspects of America's oil and natural gas industry and develops standards and recommended practices. API-650 is the most widely recognized standard for petroleum storage tanks in the world. The standard has been adopted by reference in NFPA-30 and many state fire codes. API-620 covers the design and construction of large, welded, low pressure carbon steel aboveground storage tanks.

Underwriters Laboratories (UL) is a not-for-profit organization whose mission is to establish, maintain, and operate laboratories for the examination and testing of devices, systems, and materials to determine their relation to hazards to life and property; and to ascertain, define, and publish standards, classifications, and specifications for materials, devices, products, equipment, construction, methods, and systems affecting such hazards. UL develops and maintains numerous standards related to flammable and combustible liquids storage.

The Steel Tank Institute is a not-for-profit organization that works with tank manufacturers, users, regulatory authorities, and consultants to promulgate standards for the design, construction, and installation of aboveground and underground tanks used for the storage of flammable and combustible liquids. The STI publishes consensus standards that guide tank manufacturers and code enforcement officials.

Regulations are laws developed by government agencies. At the federal level, the U.S. Environmental Protection Agency (EPA) and the U.S. Occupational Safety and Health Administration (OSHA), and the primary agencies, regulate aboveground storage tanks and their related systems. Some of the larger petroleum companies have internal engineering standards and guides that may be more stringent than other established voluntary consensus standards.

## References and Suggested Readings

1. Benedetti, Robert P., *Flammable and Combustible Liquids Handbook* (5th edition). Quincy, MA: National Fire Protection Association (2015).
2. DiGrado, Brian, and Gregory Thorp, *Handbook of Storage Tank Systems*. New York, NY: Eastern Hemisphere Distribution (2007).
3. *Federal Participation in the Development and Use of Voluntary Consensus Standards and in Conformity Assessment Activities,* Circular A-119. Washington, DC: Office of Management and Budget, Executive Office of the President (January 27, 2016).
4. National Fire Protection Association, *NFPA 1—Fire Code* Handbook. Quincy, MA: National Fire Protection Association (2015).
5. National Fire Protection Association, *NFPA 30—Flammable and Combustible Liquids Code*. Quincy, MA: National Fire Protection Association (2016).
6. Pullarcot, Sunil, *Aboveground Storage Tanks: Practical Guide to Construction, Inspection, and Testing*. Boca Raton, FL: CRC Press (2015).

# Tank Design and Construction Features

CHAPTER 3

Courtesy of William T. Hand.

## Chapter Outline

- Objectives
- Key Terms
- Introduction
- Cone (Fixed) Roof Tanks
- Open (External) Floating Roof Tanks
- Covered (Internal) Floating Roof Tanks
- Low Pressure Storage Tanks
- Horizontal Low Pressure Storage Tanks
- Pressure Vessels
- Product Transfer and Movement to/from Distribution Facilities
- Storage Tank and Distribution Facility Safety Features
- Ancillary Equipment
- Summary
- References

## Objectives

1. Given examples of the following atmospheric pressure bulk liquid storage tanks, describe the basic design and construction features and types of products commonly found for each (NFPA 472—16.2.1.3, 16.2.1.4):
   - Cone (fixed) roof tank
   - Open (external) floating roof tank
   - Open floating roof with a geodesic dome external roof
   - Covered (internal) floating roof tank
   - Vertical and horizontal low pressure storage tank
2. Describe the design and purpose of each of the following storage tank components, where present (NFPA 472—16.2.1.5):
   - Tank shell material of construction
   - Type of roof and material of construction
   - Primary and secondary roof seals (as applicable)
   - Incident venting and pressure relief devices
   - Tank valves
   - Tank gauging devices
   - Tank overfill device
   - Secondary containment methods (as applicable)
   - Transfer pumps (horizontal or vertical)
   - Tank piping and piping supports
   - Fixed or semifixed fire protection system
   - Vapor recovery units (VRU) and vapor combustion units (VCU)
   - Truck loading rack additive tanks
   - Loading rack product control and spill control systems
3. Given examples of different types of bulk flammable liquid storage tank facilities, identify the impact of the following fire and safety features on the behavior of the products during an incident, when present (NFPA 472—16.2.2.1):
   - Tank spacing
   - Product spillage and control (impoundment and diking)
   - Tank venting and pressure relief systems
   - Transfer and product movement capabilities
   - Monitoring and detection systems
   - Fire protection systems
4. Identify and describe the procedures for the normal movement and transfer of product(s) into and out of the facility and storage tanks (NFPA 472—16.2.1.2).

## Key Terms

**Cone (Fixed) Roof Tank** A sealed container with a fixed bottom and a fixed roof. Commonly constructed of carbon steel and designed for low internal pressures.

**Open (External) Floating Roof Tank** A tank shell with a fixed bottom and a floating roof. The roof moves up or down inside the tank shell with the product level.

Covered (Internal) Floating Roof Tank A combination cone roof tank with a weak roof-to-shell seam and an internal floating roof.

Vertical Low Pressure Storage Tank Cylindrically shaped tank with a fixed top and bottom, and will include a pressure/vacuum device. The roof may have either a flat or dome shape (i.e., dome roof tanks) and it is limited to a vapor pressure of 15 psig and less.

Horizontal Low Pressure Storage Tank Cylindrically shaped tank with fixed ends; commonly found in welded steel plate construction but older bolted tanks may also be found. Should be supported on protected structural supports.

Pressure/Vacuum Valves Mounted on fixed roof tanks to provide pressure/vacuum relief venting capacity for normal transfer operations.

## Introduction

This chapter will review the basic design and construction features of flammable and combustible liquid storage tanks. While there are a wide variety of design standards for tanks, they generally fall into five broad types of tanks. These include:

- Fixed (cone and dome) roof tanks
- Open (external) floating roof tanks
- Covered (internal) floating roof tanks
- Vertical low pressure storage tanks
- Horizontal low pressure storage tanks

The type of storage tank used to store flammable and combustible liquids is determined by the physical characteristics of the product being stored and the location of the tank (e.g., tank farm vs. industrial facility).

Typically, combustible liquids with flash points greater than 100°F (37.8°C) are stored in large fixed roof tanks or smaller low pressure vertical or horizontal tanks. The advantage of storing combustible liquids in these types of tanks is that they are of relatively simple construction and are less expensive to build. The disadvantage is that there is a vapor space above the liquid level, which can produce significant product evaporative losses and environmental emissions. The larger the vapor space, the greater the risk of a fire or explosion. Of course, the higher the flash point, the lower the vapor pressure and risk of ignition. Consequently, low vapor pressure liquids stored in tanks with a vapor space are viewed as an acceptable risk. If a tank fails or catches fire, there are systems in place to contain any released product and minimize the impact (e.g., dikes, berms, fixed or semifixed fire protection systems).

Products classified as flammable liquids with a flash point less than 100°F (37.8°C) are usually stored in open top or covered floating roof tanks in bulk quantities. The advantage is that there is normally no vapor space in the tank because the roof floats on the surface of the liquid.

Products with flash points less than 100°F (37.8°C) may also be stored in smaller low pressure vertical or horizontal tanks. However, when these tanks are used to store flammable liquids, they usually are limited to small petroleum marketing facilities or for gasoline additives at distribution terminals. When found in chemical manufacturing or pharmaceutical industries, some tanks storing flammable liquids may have a nitrogen "cap" to provide an inert vapor atmosphere and reduce the risk of ignition.

## Cone (Fixed) Roof Tanks

### Design and Construction Features

A cone roof tank is simply a sealed container with a fixed bottom and a fixed roof. These tanks are commonly constructed of carbon steel and are designed for low internal pressures. Fixed roof tanks designed in accordance with API-650 can be designed for pressures up to 2.5 psig (129 mmHg).

Fixed roof tanks are easily identified by their smooth welded exterior shell, although some older tanks with riveted or bolted seams may occasionally be found. The tank exterior generally lacks the wind girder and tank vents found on floating roof tanks. Insulated and heated cone roof tanks may also be found in some facilities which handle heavier, more viscous products such as asphalt and bunker oils.

Despite the name "cone roof" the roof may not actually appear to have a cone shape, especially from ground level views. The actual slope of the roof depends on the size of the tank. Large tanks with a fixed roof may have a flat or cone-shape roof, while smaller diameter tanks can be found with a dome-shape or flat roof. Depending upon user requirements, fixed roof tanks over 50 feet (15.2 m) in diameter are often column-supported, while self-supporting dome or cone roof tanks are an option for tanks under 50 feet (15.2 m) in diameter. See FIGURE 3-1.

All fixed roof tanks will have a vapor space between the liquid level and the roof. If this space is in the explosive range at the time an ignition source is introduced, an explosion will occur. Primary ignition sources for cone roof tank fires have included lightning, static electricity, and the improper use of welding and cutting equipment.

### Venting

Venting is an important safety feature. As liquid enters the internal tank space, the rising liquid compresses the vapors, which then either escape to the outside atmosphere or are

# Scan Sheet 3-A—NFPA-30 Classification of Flammable and Combustible Liquids

The definitions for flammable and combustible liquids as adopted in NFPA-30 are built around the definitions of flash point, boiling point and vapor pressure. Understanding these definitions is an important part of interpreting the various classifications and evaluating incident-specific risks.

**FLASH POINT**. Minimum temperature at which a liquid gives off enough vapors that will ignite and flash-over but will not continue to burn without the addition of more heat. This is significant in determining the temperature at which the vapors from a flammable liquid are readily available and may ignite.

**BOILING POINT**. The temperature at which a liquid changes its phase to a vapor or gas. The temperature where the vapor pressure of the liquid equals atmospheric pressure.

**VAPOR PRESSURE**. The pressure exerted by the vapor within the container. Vapor pressure is temperature dependent; as the temperature increases, so does the vapor pressure. In turn, this can potentially increase the size of the hazardous environment and the risks of ignition.

Flash point, boiling point, and vapor pressure are significant properties for evaluating the flammability of a liquid, as all three are directly related. A liquid with a low flash point will also have a low boiling point, which translates into a higher vapor pressure and a larger amount of vapors being given off. From a risk perspective, there is a higher probability for ignition of low flash point, high vapor pressure products than there is for higher flash point, low vapor pressure products.

**Examples of common flammable liquids are listed in the table.**

| PRODUCT | FLASH POINT | BOILING POINT | VAPOR PRESSURE |
|---|---|---|---|
| Gasoline | −45°F (−42.8°C) | 100°F (37.8°C) to 400°F (204.4°C) | Varies |
| Kerosene | 100°F (37.8°C) to 162°F (72.2°C) | 304°F (151°C) to 574°F (301°C) | Varies |
| Methanol | 54°F (12.2°C) | 147°F (63.9°C) | 92 mm |
| Ethanol | 55°F (12.8°C) | 173°F (78.3°C) | 43 mm |
| Toluene | 40°F (4.4°C) | 231°F (110.5°C) | 20 mm |
| Xylene | 90°F (32.2°C) | 292°F (144.4°C) | 9 mm |

**FLAMMABLE LIQUID.** A liquid having a closed cup flash point below 100°F (38°C) and having a vapor pressure not exceeding 40 psi (2,068 mmHg) at 100°F (37.8°C). Flammable liquids are organized into classes. Class 1 flammable liquids are further subdivided into three groups:

- **Class IA** includes liquids with flash points below 73°F (22°C) and having a boiling point *below* 100°F (37.8°C).
- **Class IB** includes liquids with flash points below 73°F (22°C) and having a boiling point *above* 100°F (37.8°C).
- **Class IC** includes liquids with flash points at or above 73°F (22°C) and *below* 100°F (37.8°C).

**COMBUSTIBLE LIQUID.** A liquid having a closed cup flash point at or above 100°F (37.8°C). Combustible liquids (i.e., Class II and III) are subdivided into three groups:

- **Class II** liquids include those having flash points at or above 100°F (37.8°C) and below 140°F (60°C).
- **Class IIIA** includes liquids having a flash point at or above 140°F (60°C) and below 200°F (93°C).
- **Class IIIB** includes liquids having a flash point at or above 200°F (93°C).

**FIGURE 3-1** Cone Roof Storage Tank
Courtesy of Gregory G. Noll.

**FIGURE 3-2** Pressure/vacuum (PV) valves are used for normal tank venting.
Courtesy of Gregory G. Noll.

drawn to a vapor recovery unit (VRU) or vapor combustion unit (VCU). Likewise, when the liquid in the tank is drawn down, it draws a vacuum. Venting at the top of the tank allows air to enter the tank, thereby equalizing the pressure and preventing a vacuum.

Fixed roof tanks may be equipped with open vents or a pressure/vacuum (PV) relief device mounted on the fixed roof to provide sufficient venting capacity for normal transfer operations. API 2000: Venting Atmospheric and Low-Pressure Storage Tanks describes the venting requirements for fixed roof tanks for both normal and emergency venting conditions. Open-type vents may be found for products with a vapor pressure less than 1.5 psia (77.6 mmHg). For higher vapor pressure products, PV devices are commonly installed.

Historically, PV valves have been the source of many fires on cone roof tanks. Tanks being filled are also exhausting flammable vapors which can travel to adjacent ignition sources. For example, hot work operations have been responsible for starting several vent fires. Another common source of vent fires is lightning. Consequently, bulk petroleum facilities may suspend filling operations during severe lightning storms.

It is important to recognize that PV valves are designed for normal operations and are not designed as an emergency pressure relief system (**FIGURE 3-2**). The presence of the valve does not in any way mean that the roof may not fail violently if the tank has direct flame impingement. For example, a large dike fire or flame impingement in the vapor space from an adjacent ground or tank fire could cause the fixed roof to fail violently. The belief that the PV valve would relieve the internal pressure during firefighting operations was a key factor in the death of three firefighters at a 1976 fire in Gadsden, Alabama. See page 00 for a more detailed review of this incident.

## Flame Arresters

Flame arresters may be found on cone roof storage tank PV valves. The purpose of a flame arrester is to stop flame propagation from passing from outside the tank, through the valve, and into the interior vapor space. However, clogging from dirt and insect nests, icing from cold weather, and other situations may prevent the flame arrester from working properly. If flame arresters are to be effective, they cannot be over 6 pipe diameters from the tank shell.

## Weak Roof-to-Shell Seam

To minimize the potential for violent tank rupture during overpressure scenarios, a major feature of a cone roof tank design is the weak roof-to-shell seam. The tank roof is engineered to prevent tanks from separating at their base and rocketing. While not common, there have been several serious incidents where smaller diameter fixed roof tanks have violently separated at their base and were blown skyward very similar to a rocket launch. In most cases, these tanks were constructed to petroleum industry standards. See **FIGURE 3-3 and 3-4**.

Tanks constructed in compliance with API-650 have a 3/8-inch weld seam connecting the roof to the shell, which is designed to yield to the force of an explosion in the vapor space. While this is an important safety feature, don't get the impression that roof separation is like popping the top on a soda can. When the fixed roof on a 100-foot-diameter cone roof storage tank separates, the weak roof to shell seam may fail in a number of different manners and behaviors. Failure modes may include a portion of the roof peeling back, tearing of the tank shell along the roof to shell seam (i.e., "fish mouth"

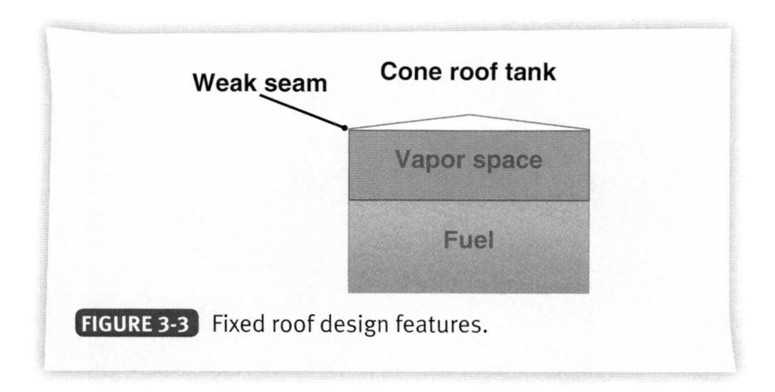


FIGURE 3-3 Fixed roof design features.

FIGURE 3-4 Fixed roof separation.

Courtesy of John Hess.

opening), or parts of the roof tearing away and fragments either falling into the tank or impacting surrounding tanks and exposures. Partial tears will present challenges to emergency responders in applying foam in sufficient quantities onto the surface area.

## Open (External) Floating Roof Tanks

### Design and Construction Features

An open top floating roof tank is essentially a tank shell with a fixed bottom and a floating roof. The roof moves up or down inside the tank shell with the product level. The main advantage of the floating roof over the fixed roof is that, with the exception of maintenance operations when the roof is in the low position, there is no vapor space between the liquid and the roof.

Open top floating roof tanks can be distinguished by the characteristic wind girder that rings the top of the tank. This ring acts as a stiffener for the top of the tank shell, giving it structural support when the roof is in the lower position inside the tank. Depending upon tank construction features, the wind girder is typically not

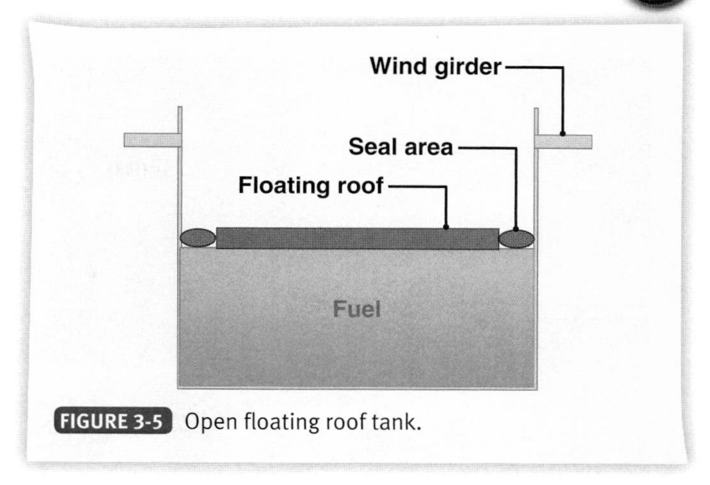


FIGURE 3-5 Open floating roof tank.

designed to serve as a walking surface to inspect the tank. See FIGURES 3-5, 3-6, and 3-7.

Design requirements for external floating roof tanks can be found in API-650, Appendix C.

### Floating Roof Types

The roof on an external floating roof tank literally floats on top of the liquid. This is possible because air is trapped in small compartments in the roof much like the flotation compartment on small boats that allow the boat to float even if it is swamped or flooded.

There are two basic styles of floating roofs found on external floating roof tanks: pontoon or double-deck designs. The most common style is the pontoon roof, which is supported by a compartmented annular ring of pontoons and a center single deck; the pontoons are welded and sealed to prevent air from escaping and rainwater or product from entering. Pontoon roofs are designed so that the roof can still float even if several pontoon compartments fail due to leakage or rupture during exposure to fire. At some point, however, the roof will sink if enough pontoon compartments are breached.

The second common floating roof design found on external floating roof tanks is the double-deck floating roof, where two complete decks are joined by a series of concentric rings. The outer annular bay is compartmented to provide buoyancy, and emergency overflow drains are provided to prevent stormwater from exceeding the capacity of the roof.

A third design sometimes found is the "honeycombed" roof. This type of roof incorporates flotation into the roof itself and provides flotation stability. This design incorporates several hundred small honeycombed compartments which are sandwiched between the top and bottom of the floating roof. These roofs are generally more expensive to install, but they have a good performance record when exposed to fire because the roof is not subject to the same type of failure as when several small cells fail. In other words, the weight of the roof is distributed among many small air cells rather than just a few cells as is the case with the pontoon roof.

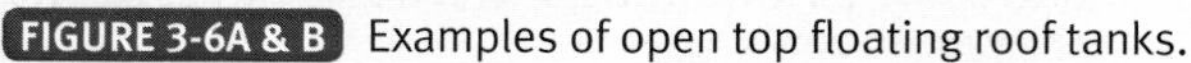

FIGURE 3-6A & B Examples of open top floating roof tanks.

Courtesy of Gregory G. Noll.

FIGURE 3-7 Floating roof tanks are confined spaces when the roof is 8 feet below the top.

Courtesy of Tyler Bones.

Both the pontoon and floating double-deck style roofs are designed to carry a normal "live load" plus additional loads that may be created by rainwater or snow accumulation. Floating roofs have drainage systems designed to carry normal rainwater off the roof and to the ground where it is collected inside the diked area. At some point, however, the weight of an unusual water load or snow pack can partially or completely sink the roof. When this happens the product inside the tank is exposed to the atmosphere and vapors are subject to ignition.

A floating roof may also sink due to the additional weight created by fire streams especially if the roof is already compromised (e.g., not level) at the beginning of the incident. For example, if fire water streams are arbitrarily directed over the top of the tank to control a seal fire, the drainage system normally cannot carry the water away fast enough. When this condition exists, the roof can tilt to one side and either sink partially or completely. This problem is discussed in more detail in Chapter 6. Obviously sinking the roof is a very hazardous condition. It doesn't do much for career advancement either.

The cutaway view in FIGURE 3-8 shows an open top floating roof with a pontoon float around the roof pan. Note the pipe legs supporting the roof of the tank. These legs are raised (i.e., short leg position) for normal operation and lowered (i.e., long leg position) to support the roof during periods of maintenance. During tank maintenance operations, this area below the roof should be treated as a confined space. See FIGURE 3-9.

## Floating Roof Seals

As the liquid level in the tank changes, the floating roof adjusts its position inside the tank shell by sliding or "floating" up and down with the level of the tank. This vertical movement is made possible by seals which are installed between the tank shell and the floating roof. Depending upon tank diameter, the seal area may be 1 to 4 feet (0.3 to 1.2 meters) wide. The seals serve two purposes. First, they provide room between the shell and the roof to slide up and down without touching the tank shell. Second, the seal serves as a barrier between the product and the atmosphere, and reduces evaporative emissions. Basically, the seal keeps water out and vapors in.

Two types of seals are commonly found. These include the fabric seal with pantograph and the tube seal (see FIGURE 3-10). Environmental regulations require that floating roof seals prevent emissions to the outside atmosphere; this is often accomplished through the use of a double shoe seal design. Properly designed and maintained

FIGURE 3-8 Floating roof tank construction features.
© Jones & Bartett Learning.

FIGURE 3-9 Interior view of a floating roof tank.
Courtesy of Gregory G. Noll.

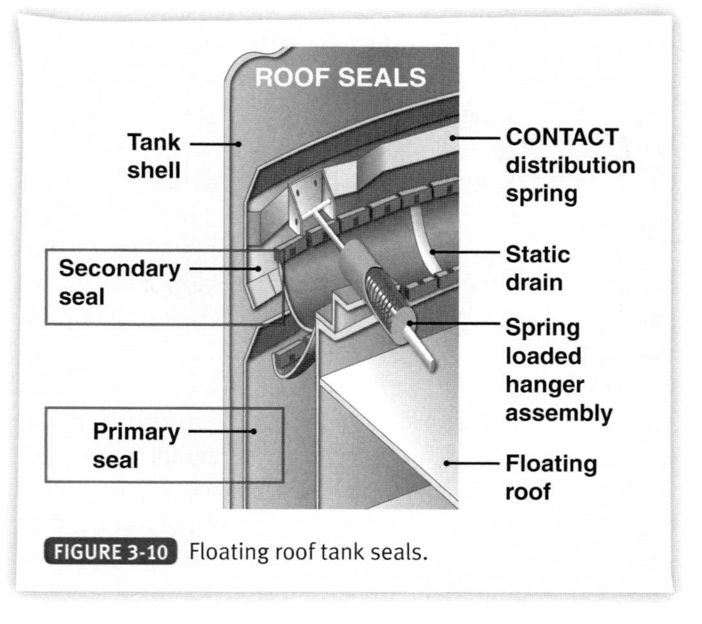


FIGURE 3-10 Floating roof tank seals.

seals and tanks can reduce evaporative and environmental emissions by over 98%.

## Shunts and Lightning Protection

Many petroleum products such as gasoline are static charge accumulators. This means that the liquid picks up a static charge much the same as a person can accumulate a static charge when walking across a carpet. The resulting static spark can contain sufficient energy to ignite flammable vapors.

Static accumulation can occur with rapid loading rates of product into the storage tank. This problem can be avoided by slowing down the loading rate of large tanks by following published industry tables. Slower loading rates ensure that there is a downstream relaxation time which allows the static charge to dissipate to a safe level before it reaches the tank. A likely place for a static discharge in a floating roof tank is between the tank shell and the roof across the space filled by the seal.

The height of aboveground petroleum storage tanks also makes them good targets for lightning strikes. When a tank is struck by lightning the shortest pathway to the ground is through the metal shell. If the roof is struck by the lightning bolt, the electrical energy may jump from the floating roof to the tank shell causing a fire in the seal area.

To reduce the risks associated with static electricity and lightning strikes, the seals between the floating roof and the tank shell will be equipped with metal shunts, which provide contact between the roof and the tank shell as the roof moves up and down the tank. Based upon API RP 45: Recommended Practice for Lightning Protection of Aboveground Storage Tanks for Flammable or Combustible Liquids, the shunts should be submerged by 1 foot (3 meters) or more and installed at 10-foot (3-meter) intervals around the roof perimeter to provide a conductive pathway from the tank roof to the tank shell in the event of a lightning strike.

## Covered (Internal) Floating Roof Tanks

### Design and Construction Features

The covered floating roof tank (also known as an internal floating roof tank) includes the same basic construction features of the external floating roof tank except that there is a fixed roof at the top of the tank. Essentially, it is a combination cone roof tank with a weak roof-to-shell seam and an internal floating roof. Some tanks may utilize a domed roof design rather than the traditional coned roof. See FIGURE 3-11.

FIGURE 3-11 Cone roof tank compared (top) to an internal floating roof tank (bottom).

Courtesy of Gregory G. Noll.

Covered floating roof tanks can be identified and distinguished between the cone roof tank and the open top floating roof tank by their characteristic "eyebrow" vents at the top of the tank shell. These vents allow air to escape and enter the inside space between the cone roof and the internal floating roof as it moves up and down the tank.

The primary advantages of the covered floating roof tank over the open top floating roof tank include reduced environmental emissions (98% and higher) and not having the floating roof deck exposed to the elements. The covered floating roof tank also has an excellent safety record, with very few of the fire problems that the cone roof tank and open top floating roof tank have historically experienced. Given these advantages, most newly constructed tanks at petroleum manufacturing and distribution facilities are of the internal floating roof design. The downside from an emergency response perspective is that fires involving this design can be difficult to extinguish.

The covered floating roof tank construction features include vapor space vents at the top of the tank, cone roof support legs that pass through the floating roof, and pipe legs that act as stops for the floating roof when it is resting on the bottom of the tank. See FIGURE 3-12.

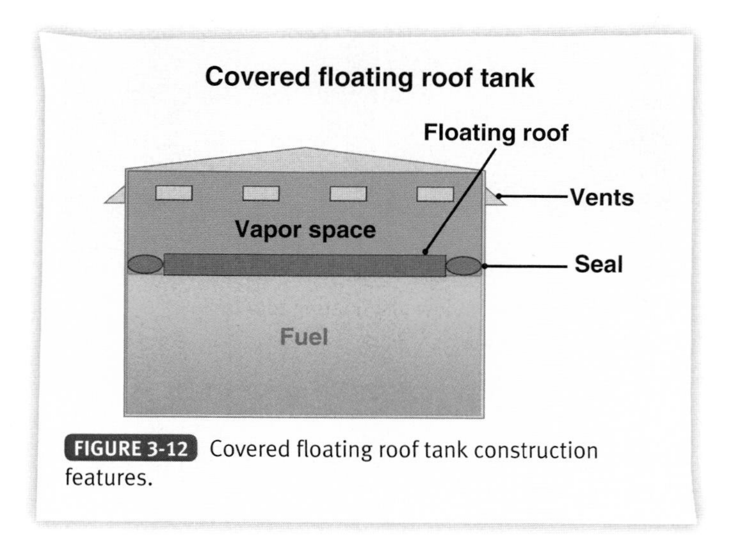


FIGURE 3-12 Covered floating roof tank construction features.

A hybrid of the covered floating roof design is the installation of an aluminum geodesic dome over an existing external floating roof tank. This design has been in existence for many years, especially in the northern latitudes of the United States. From an operational perspective, the use of an aluminum geodesic dome allows a facility operator to convert an open floating roof tank into a "modified" covered roof tank with the benefits of reduced cost, product and environmental emissions, and weather (i.e., snow, rain) impacts. See FIGURE 3-13.

FIGURE 3-13 Open floating roof tanks with geodesic dome.

Courtesy of Gregory G. Noll.

## Floating Roof Types

The most common type of floating roof found on an internal floating roof tank is a single deck pan (commonly referred to as the internal pan) constructed of welded steel or aluminum. However, depending upon the facility (refinery, distribution terminal, etc.), the pontoon, double-deck, or honeycomb design may also be found. See FIGURE 3-14.

Day-to-day operation and performance of the floating roof on an external floating roof tank is comparable to what is found on an internal floating roof tank, with one design exception. In lieu of the internal roof being supported on pipe legs when in the low position, a design in which an aluminum internal roof can be suspended by cables attached to the underside of the fixed roof or aluminum dome may be found. Facility operators cite reduced emissions, maintenance costs, and maintenance flexibility with the use of the cable suspension roof design. See FIGURE 3-15.

One of the challenges in dealing with the internal floating roof tank in a fire scenario will be the integrity and position of the internal floating roof. Incidents have included scenarios where the internal roof partially sinks or creates spaces, thereby creating void spaces where it may be difficult to apply foam streams onto the fuel surface.

FIGURE 3-14 Examples of covered floating roof tanks.

Courtesy of Gregory G. Noll.

FIGURE 3-15 Cable suspended aluminum roof.

Courtesy of Allentech, Inc.

FIGURE 3-16 Vertical welded storage tank. Note the external thermal fire protection on the tank shell.

Courtesy of Gregory G. Noll.

## Low Pressure Storage Tanks

### Design and Construction Features

Low pressure vertical storage tanks have relatively simple features. They are cylindrically shaped with a fixed top and bottom, and will include a PV device. The roof may have either a flat or dome shape (i.e., dome roof tanks). Depending upon the type of facility and product service, some vertical storage tanks may also be insulated (see FIGURE 3-16).

Low pressure vertical storage tanks can be divided into two broad categories. These include bolted and welded construction.

- **Bolted vertical tanks** include the API-12B tanks of capacities ranging from 100 to 10,000 barrels (4,200 to 420,000 gallons). These are basically a small cylindrical tank constructed from bolted steel plate with a PV valve. Most 12B tanks are found in crude production fields; however, there are many older bolted tanks meeting similar specifications in other types of service. API-12B tanks are usually sited on a concrete pad.
- **Welded vertical tanks** include the API-12D tanks of 500 to 10,000 barrels capacity, the API-12F tanks of 90 to 500 barrels up to a maximum of 16 feet (4.87 meters) in diameter, and larger.

  Vertical tanks built to API-620 specifications are limited to a vapor pressure of 15 psig and less. They must be protected by automatic pressure-relieving devices that will prevent the pressure at the top of the tank from rising more than 10% above the maximum allowable working pressure (MAWP).

API 2000: Venting Atmospheric and Low-Pressure Storage Tanks describes the venting requirements for low pressure tanks for both normal and emergency venting conditions. Although not common, vertical tanks not designed with a weak roof-to-shell seam may fail at the bottom seam in a fire scenario and then rocket.

FIGURE 3-17 Vertical storage tanks can often be found at distribution terminals for the storage of fuel additives, dyes and detergents.

Courtesy of Gregory G. Noll.

Smaller vertical tanks may also be found at petroleum distribution facilities for the storage of fuel additives, dyes, detergents, etc (see FIGURE 3-17). These tanks are usually found in proximity to the cargo tank truck loading racks, and their products are blended into gasoline at the loading rack, which are blended into the product (see FIGURE 3-18).

FIGURE 3-18 Gasoline additive tanks at a cargo tank truck loading rack.

Courtesy of Gregory G. Noll.

## Horizontal Low Pressure Storage Tanks

### Design and Construction Features

A wide range of horizontal storage tanks can be found in petroleum service. Like their vertical cousins, horizontal tanks may be found in either bolted or welded steel plate construction. Historically, the bolted horizontal storage tank has a bad reputation in the fire service because it often fails rapidly when subjected to fire. These types of tanks are especially a problem when constructed with unprotected structural steel supports. In some situations involving older tanks, fire investigations have shown that the product stored in a horizontal tank at the time of an incident had different physical and chemical properties when compared to those for which the tank was originally constructed (e.g., higher vapor pressure). As a result, the pressure relief device was not adequately sized for the change in product storage and the container failed more rapidly.

Most modern horizontal storage tanks are built to meet UL Standard 142 requirements. UL 142 tanks are cylindrical in shape and are constructed, inspected, and tested for leakage before being shipped from the factory as completely assembled vessels (see FIGURE 3-19). They are used primarily for storing flammable and combustible liquids at pressures in vapor spaces between atmospheric and 0.5 psig. They are commonly found in service at businesses that do not need large storage capacities and at installations that meet the requirements of the NFPA 30: Flammable and Combustible Liquids Code.

These new-generation horizontal tanks have an excellent fire record. These tanks are often insulated and have an integrated spill confinement system built around the storage tank.

FIGURE 3-19 Horizontal storage tank configurations.

Courtesy of Gregory G. Noll.

### Structural Supports

Most fire codes require that horizontal tanks be installed on firm foundations such as concrete, masonry, or protected steel. Under current editions of most fire codes, unprotected steel supports are prohibited. Experience has shown that they can soften and sag when exposed to fire and can fail after only a short period of time. This risk became known nationally after the 1959 storage tank incident that occurred in Kansas City, Kansas. Six firefighters were killed and numerous others injured when a burning horizontal tank failed after the pressure relief valve functioned. When the tank failed, the flammable product engulfed firefighters who were using hose streams off the end of the tank.

Steel supports for tanks containing Class I, Class II, or Class IIA liquids must be protected by materials having a fire resistance of not less than 2 hours. Some jurisdictions also require that steel supports on elevated horizontal tanks be protected by water spray systems.

## Pressure Vessels

As noted in the introduction, the scope of this text does not include high pressure vessels such as bullet tanks and spheres found in liquefied petroleum gas (i.e., propane and butane) service. They have been excluded as a way to limit the topic and scope of the book.

However, pressure vessels are often found either on or in close proximity to flammable liquid manufacturing, storage, and distribution facilities (see FIGURE 3-20). The risks of a boiling liquid, expanding vapor explosion (BLEVE) are well documented within the emergency response community, and have been responsible for the deaths of numerous emergency responders from a historical perspective.

NFPA 30 **does permit** pressure vessels to be used as low pressure storage tanks. Vessels used in these situations were originally designed and used as pressure vessels and were converted for use as low pressure storage containers. The normal operating pressure of the vessel must not exceed the design pressure of the vessel.

FIGURE 3-20 LPG pressure vessels adjoining a bulk petroleum storage facility.

Courtesy of Gregory G. Noll.

## Product Transfer and Movement To/From Distribution Facilities

Petroleum products may be transferred into and out of petroleum manufacturing, storage, and distribution facilities by pipeline, marine, rail, and cargo tank truck modes. For petroleum distribution facilities, the most common transportation mode for moving petroleum products into the facility is by pipeline, whereas ethanol is usually moved by rail or cargo tank truck. This section will focus on pipelines, and loading racks will be discussed later in this section.

Pipelines and piping systems are the safest and second largest hazmat transportation mode within the United States. From a design and construction viewpoint, all piping systems are based upon the following principles:

1. *A material is inserted or injected into a pipe.*
2. *The product is moved from this origination point to a prespecified destination.* Flammable liquid products are physically moved as a result of energy created through the use of pumps. In addition, various valves and manifolds may be used to control and direct the flow of the product.
3. *The product is ultimately removed from the pipeline at its destination point.* Depending upon the type of pipeline and the location, the product may be transferred to another mode of transportation (e.g., marine, rail, highway), placed into a container for storage (e.g., tank), or used.

Pipelines often cross over or under roads, waterways, and railroads in close proximity to the storage tank facility. At each of these crossover locations, a marker should identify the pipeline right-of-way. Although its format and design may vary, all markers are required to provide the pipeline contents (e.g., natural gas, propane, liquid petroleum products), the pipeline operator, and an emergency telephone number (see FIGURE 3-21).

The pipeline emergency telephone number goes to a control room, where an operator monitors pipeline operations and can start emergency shutdown procedures. It should be stressed that even when a ruptured pipeline is immediately shut down, product backflow may continue for several hours until the product drains to the point of release.

For liquid petroleum pipelines, there is usually no physical separator (e.g., sphere or pig) between different products. Rather, the products are allowed to "co-mingle." This interface can range from a few barrels to several hundred, depending on the pipeline size and products involved. Verification of the shipment arrival at the storage facility is made by examining a sample of the incoming batch for color, appearance, and/or chemical characteristics (see FIGURE 3-22).

**FIGURE 3-21** Pipeline markers must provide the pipeline contents, the pipeline operator, and an emergency phone number.

Courtesy of Gregory G. Noll.

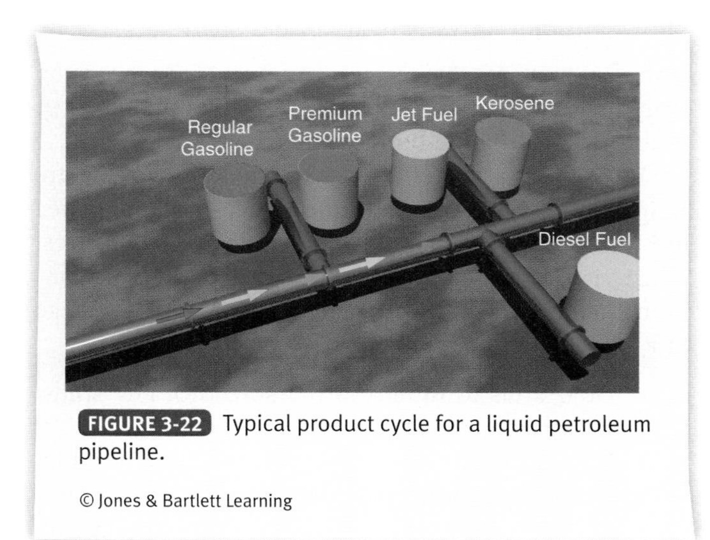


**FIGURE 3-22** Typical product cycle for a liquid petroleum pipeline.

© Jones & Bartlett Learning

Product flows through many transmission pipeline systems are monitored through a computerized pipeline Supervisory Control and Data Acquisition (SCADA) System. The exact injection date and time of the particular product into the pipeline is noted and its delivery date/time is projected. As the product gets close to its destination, a sensor in the line signals the arrival of the shipment. The SCADA System provides pipeline personnel with the ability to monitor pipeline flows and pressures and initiate emergency shutdown procedures in the event of a release.

Emergency response personnel should NEVER attempt to isolate any pipeline valves on transmission or distribution pipelines unless under the direction of pipeline operations personnel. Operation of pipeline valves and systems must be under the full command and direction of the pipeline operator. Failure to follow this injunction may create additional problems that are far worse than the original event (see **FIGURE 3-23**).

**FIGURE 3-23** Emergency responders should never attempt to isolate any pipeline valves on transmission or distribution pipelines unless under the direction of pipeline operations personnel.

Courtesy of Gregory G. Noll.

Readers may reference the *Pipeline Emergencies* (3rd edition) textbook and related Internet-based training curriculum at www.pipelineemergencies.com for additional information. The curriculum was a result of a cooperative agreement between the U.S. Department of Transportation's Pipeline and Hazardous Materials Safety Administration (PHMSA) and the National Association of State Fire Marshals (NASFM).

## Storage Tank and Distribution Facility Safety Features

The safe operation of liquid petroleum storage facilities is addressed through a number of regulations and industry standards, including NFPA 30: Flammable and Combustible Liquids Code and American Petroleum Institute (API) standards. These "standards of care" address physical, environmental, safety, and health issues, and should be factored into the risk analyses during both the planning and response phases. In this section, environmental and fire safety issues related to petroleum storage operations will be reviewed.

### Facility Layout/Tank Spacing

Bulk storage tank operations require moving large volumes of flammable liquids to facilitate storage, pipeline deliveries, and distribution operations. Emergency responders conducting a walk-through of a bulk storage facility should trace the flow of product both into and out of the facility.

From an emergency planning perspective, bulk petroleum storage tank facilities can be broken into the following elements:

- **Product entry and transfer *into* the facility.** Regardless of the mode of transportation that brings the product to the facility (e.g., marine, pipeline, rail), product is transferred into the facility through a series of valves and piping to its assigned storage tank.

- **Storage tank farm**
- **Product transfer *out* of the facility.** At most liquid petroleum distribution facilities, product leaves the facility through cargo tank trucks, while refineries may ship their products via pipeline, marine, or rail.

**Tank Spacing.** Storage tank fires can generate a significant amount of radiant heat and expose nearby tanks to excessive heating, thereby causing damage to unprotected structural steel. To decrease the risk of fire spread from tank-to-tank, adequate spacing between tanks inside a common dike area is required. Tank-to-tank and tank-to-property line spacing is based on several factors, including (1) type of tank (i.e., cone, open floater, covered floater), (2) product contained (i.e., Class 1, 2, or 3 liquid), (3) minimum distance from the tank to the property line that can be built upon, and (4) minimum distance from the nearest side of any public right of way or the nearest important building on the same property. The key point is that the further away tanks are from each other and other exposures like buildings, the less the risk of the fire spreading from one tank to another. Tank spacing should also consider access for firefighting apparatus and equipment. See **FIGURE 3-24**.

## Product Control Options and Considerations

Causes of product releases at bulk petroleum storage facilities include valve and piping leaks, tank overfills, and natural events such as flooding and earthquakes. While exceptionally rare, there have also been instances of catastrophic tank shell failure. An example of this was the January 1988 total collapse of a 4-million-gallon diesel storage tank in Floreffe, Pennsylvania. Approximately 1 million gallons of product immediately splashed over the dikes, flowed into a storm drain, and subsequently ran into the Monongahela River. The spill temporarily contaminated drinking water for over 1 million people for several days down river starting in Pittsburgh, where the Monongahela and Allegheny rivers converge, and later along the Ohio River valley. Millions of fish were also killed as a result of the spill.

**FIGURE 3-24** The layout of flammable liquid bulk storage facilities is governed through a combination of government regulations, consensus standards, and industry best practices.
Courtesy Gulf Oil Limited Partnership.

Leak detection for aboveground storage tank systems can be accomplished using periodic or continuous monitoring. Monitoring refers to surveillance of the system components and/or the subsurface area to promptly detect storage tank spills or leaks.

Leak detection methods for tanks include visual inspection, inventory control, product level monitoring, subsurface monitors that monitor groundwater or soil vapor, sump monitors in dike areas, and interstitial monitors in double-wall or double-bottom tanks.

Product confinement and containment systems are designed to confine the released product to a limited area and minimize its impact upon surrounding exposures. When viewed from a systems perspective, product containment structures and options for aboveground storage tank systems include the following:

- Dike walls to minimize the spread of any spills/releases outside the immediate tank area
- Impoundment areas where the released product flows to a remote area away from the storage tank
- Double-walled tanks, which includes a storage tank within a larger, outer storage tank
- Double-bottom tanks to minimize tank leakage into the ground and groundwater
- Secondary containment, such as rubber or other impervious liners, that minimize product movement inside a dike wall into the ground and groundwater

Additional information on each of these possible options is provided next.

### Primary Containment Measures

**Dikes.** The design and construction of the dike that surrounds one or more storage tanks is based upon the size/diameter of the tank(s) being protected, the compatibility of the dike or impoundment material with the product(s) being stored, the desired quality of construction, and local code requirements.

Dikes are constructed from earth, concrete, or other impervious materials, and are intended to confine a release from a tank or its associated piping inside the diked area. The total capacity of the dike is based upon the capacity

FIGURE 3-25 Examples of storage tank dikes.

Courtesy of Gregory G. Noll.

FIGURE 3-26 Example of a storage tank dike with secondary containment.

Courtesy of Gregory G. Noll.

FIGURE 3-27 Smaller dikes may be found within a containment dike to handle smaller spills and releases.

Courtesy of Gregory G. Noll.

of the single largest tank within the dike. If there are multiple storage tanks within a common "outer" dike, there may also be smaller tank dikes around the individual storage tanks that can contain smaller spills and releases up to 15% of the capacity of the tank being protected. See FIGURE 3-25, 3-26, and 3-27.

Dikes are equipped with drains to remove accumulated rainwater. These valves are normally maintained in the closed position to prevent flammable liquid from flowing out of the dike during a spill. Failure to ensure that the dike drain was closed was a critical factor in an April 1988 incident in Martinez, California, where the roof drainage system on a 300,000-barrel (12.6-million-gallon) storage tank failed as a result of malfunctioning equipment.

According to incident reports, a flexible rainwater drain pipe designed to drain water from the tank roof separated from its coupling and a cutoff valve on the storage tank was left open by workers, thereby allowing crude oil to flow into the dike. Because the dike drain was also open, approximately 400,000 gallons of crude oil drained from the tank into a marsh and subsequently into the Carquinez Strait, a waterway that empties into San Francisco Bay. The oil spill caused extensive damage to the ecosystem, killed wildlife in the vicinity, and damaged wetlands.

During spill control and firefighting operations it is critical that all dike drains be checked to ensure they are in the closed position. When tank exposure to fire is an issue, cooling water applied to the tank's shell can accumulate in depth inside the diked area. If hydrocarbon liquid is floating on top of the spill a serious hazard can develop if the dike fails or overflows, or the facility storm and sewer system is overwhelmed. This was a critical factor in the death of eight firefighters on August 17, 1975, during a fire at the former Gulf Oil Refinery in Philadelphia, Pennsylvania.

FIGURE 3-28 Remote impoundment basin.

Courtesy of Gregory G. Noll.

**Remote Impoundment Areas.** These areas are usually constructed or graded so that any spilled product flows away from the tank and exposures by way of grading, swales, and ditches, to an area large enough to contain all of the liquid of the largest tank that can drain into it. Impoundment areas may also be constructed of concrete and may have a berm or curb to define the area. NFPA 30 requires that the remote impoundment also be separated from adjacent buildings or other facilities. See FIGURE 3-28.

**Double-Wall Tanks.** This design is simply a primary tank (i.e., interior tank) within a second tank (i.e., exterior tank), with containment provided by the exterior tank wall. This design may be found on steel and reinforced thermosetting plastic (RTP) tanks. Their capacity is the volume of the interior tank plus the available volume of the interstitial space (i.e., the air space between the two tank walls).

Double-wall steel tanks may be found in areas where there is insufficient space for a dike. Examples include marine terminals and harbor areas, hazardous waste storage areas, and airports. They are typically constructed onsite. In comparison, double-wall RTP tanks are generally manufactured at specialized facilities and then transported to the site for transportation. As a result, their size is limited to that which can be transported.

Whenever the exterior shell of a double-walled metal tank is used for leak containment, the design specification of both the inner and outer shells should be similar. For example, the exterior tank must be capable of withstanding the same pressure as the interior tank. Provisions should also be made for leak detection within the interstitial space. See FIGURE 3-29.

### Secondary Containment Measures

Secondary containment measures are spill control measures that primarily protect against environmental impacts caused by the migration of spills and leaks into the ground and groundwater. Secondary containment measures include double-bottom storage tanks and the use of synthetic liners. See FIGURE 3-30.

FIGURE 3-29 Example of a tank-within-a-tank containment.

Courtesy of Steve Hess, BWI Airport Fire & Rescue Department.

**Double-Bottom Tanks.** This design is used to provide improved control of tank corrosion, as well as leak monitoring and detection. Storage tanks already in service can also be retrofitted with a double bottom. Corrosion protection systems can be integrated into the double-bottom design, thereby reducing the probability of a tank leak through corrosion. In addition, leak monitoring systems are often used to monitor any leakage from the "inner" tank bottom into the "secondary" bottom before there are any environmental impacts.

**Synthetic Liners.** Flexible synthetic or rubber membrane liners can provide a high level of secondary containment protection and reliability. These can be installed in several different configurations, including dike walls and bottom surfaces; in underground pipe runs; and around loading racks, pumps, and transfer equipment. They are often combined with leak monitoring and detection technologies to provide leak surveillance.

While having a low potential for mechanical damage, considerations that may influence the use of synthetic liners can include poor or gaseous soil conditions, a high water table, and the potential for flex fatigue.

**Ancillary Equipment.** A well-designed containment system should also protect against leaks from both the tank and ancillary equipment located within the primary containment area. For piping, secondary containment options can include double-wall pipe, release collection trenches, and release collection troughs.

FIGURE 3-30 Secondary tank containment (rubber liner) and oil detector being installed in a storage tank dike under construction.

Courtesy of Gregory G. Noll.

FIGURE 3-31 Piping runs and pipe racks can be a significant fire exposure problem.

Courtesy of Gregory G. Noll.

Valves and fittings are sometimes fitted with shields or jackets to collect minor leaks and to provide personal protection. Double seal valves may also be used to minimize the potential for leaks.

## Ancillary Equipment

### ■ Piping

Piping is used to move product to and from storage tanks within a facility. Piping in and around storage tanks and containment areas can contribute significantly to storage tank firefighting challenges. A sealed pipe is essentially a pressure vessel and can fail violently when impinged upon in a fire. See FIGURE 3-31.

Many types of piping materials are available for use in storage tank systems. Commonly used materials include carbon steel, stainless steel, and nonmetallic pipe such as fiberglass-reinforced plastic. All pipe materials should be chemically compatible with the substance(s) being transferred. Standards such as ANSI/ASME B31.3 address corrosion control and the design and fabrication of chemical process piping.

Piping may have either welded, threaded, or flanged connections. The potential for piping leaks can be reduced by minimizing the number of pipe connections. Welded connections in metallic piping can minimize leaks as well as air emissions, but may limit operational and maintenance flexibility provided by threaded or flanged connections. In contrast, flanged connections can expand when exposed to fire and result in a three-dimensional fire scenario. Thermal expansion from the transfer of high temperature substances must be also accounted for in the design of the piping and pipe support system. However, thermal protection on piping systems for normal operations is not designed to provide thermal relief protection in a fire scenario.

Tanks and associated piping systems may also be protected from corrosion via various methods. The most common methods are paintings and coatings; however, cathodic protection may also be found. The use of a passive galvanic protection will involve the use of sacrificial anodes that deteriorate in lieu of the equipment; however, some tanks and pipes use an impressed current which uses an external DC source. It is important to understand if such impressed current systems are in use when dealing with unignited spill scenarios.

## ■ Piping Supports

Pipe runs may also rely upon support structures for their integrity. When exposed to fire, these supports may weaken and cause piping to sag and breach at the pipe flange.

Pipe support elements such an anchors, guides, and hangers must be properly designed to support the loads transmitted by piping to the attached structures and equipment. The supports are generally made of steel, although other materials such as concrete and wood can be used. Supports should be capable of absorbing pipeline stresses to prevent damage to valves and other components.

## ■ Pumps

Both the design and placement of pumps, motors, and remote shutoffs are important from a fire protection perspective. Transfer pumps are often located inside a secondary containment area to protect them from leaks caused by seal failures and minimize the spread of any released product. Pumps may also be elevated within the secondary containment area to protect them from accumulated spilled substances or rainwater. See FIGURE 3-32.

Transfer pumps may be either vertical or horizontal, and are usually electrically powered. Pumps exposed to a fire usually destroy the pump seal and contribute to the size of the fire. To reduce this risk, some pumps are designed with closed seals to provide additional thermal protection. In addition, monitoring equipment may be installed in the pump containment area to detect any liquid leaks, sound an alarm, and shut down the pump.

Pump and pump seal fires are usually easy to control, but they can contribute to an escalating fire involving other equipment. The November 1990 tank farm fire near the Denver (Colorado) Stapleton Airport originated in an operating fuel pump. The fire spread to other equipment and eventually involved the storage tanks. One of the largest storage tank losses in U.S. history, the fire loss was $40 million in 2017 dollars.

## ■ Instrumentation

Instrumentation such as level sensors and gauges are used to monitor storage tank systems and operations for a variety of purposes. Temperature, pressure, flow rates, and chemical compatibility will determine the design of the instruments. Instruments may sound a local alarm or may be remotely connected to send an alarm to a central control room.

Historically, tank overfills have been a leading cause of serious incidents in the petroleum and related process industries. The cause of both the 2005 Buncefield, United Kingdom, and the 2009 Bayamon, Puerto Rico, fires and vapor cloud explosions were tank overfill scenarios (see FIGURE 3-33). To minimize this potential risk, tank overfill prevention systems are required for monitoring liquid levels on petroleum storage tanks as they are being filled. API STD 2350: Overfill Protection for Storage Tanks in Petroleum

FIGURE 3-32 Transfer pumps are often located within a secondary containment area.

Courtesy of Gregory G. Noll.

FIGURE 3-33 The 2009 Bayamon, Puerto Rico, tank fire and explosion was caused by a tank overfill.

US Chemical Safety Board

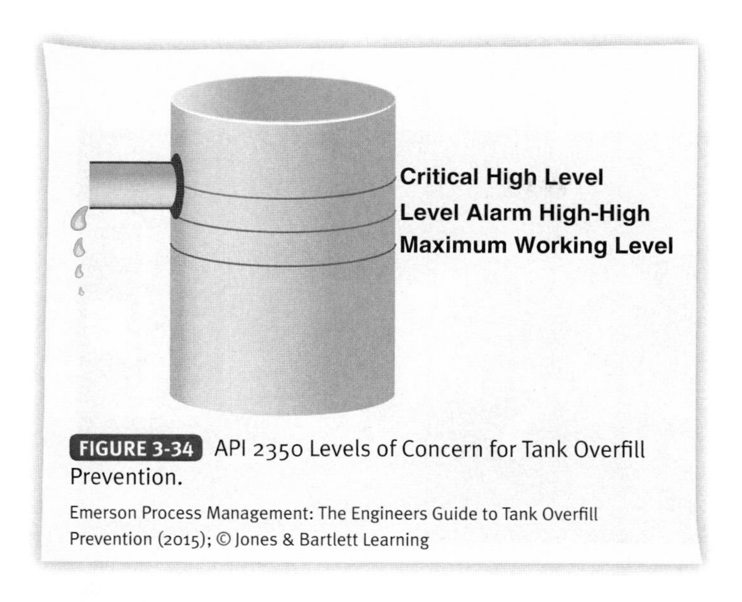


FIGURE 3-34 API 2350 Levels of Concern for Tank Overfill Prevention.

Emerson Process Management: The Engineers Guide to Tank Overfill Prevention (2015); © Jones & Bartlett Learning

FIGURE 3-35 Dike drain valves should be accessible to firefighters.

Courtesy of Gregory G. Noll.

Facilities (4th edition) identifies three levels of concern (LOC) for tank overfill prevention (see FIGURE 3-34):

- Maximum working level (MWL)—An operational level that is the highest product level to which the tank may routinely be filled during normal operations.
- Level alarm high-high (LAHH)—An alarm is generated when the product level reaches the high-high tank level. The alarm requires immediate action.
- Critical high (CH)—The highest level in the tank that product can reach without detrimental impacts (i.e., product overflow or tank damage).

Tank overfill alarms at these levels of concern will require specific actions by facility operators and/or the pipeline control room. Coordination between the receiving site and the pipeline operator is critical, as quickly shutting down a liquid transmission pipeline will generate the same hydraulic effects as a water hammer in firefighting operations. Coordinated actions can include slowing down the delivery of the pipeline shipment, diverting the liquid to another storage tank that can accept the additional volume of liquid, or shutting down the pipeline flow.

## Valves, Connections, and Couplings

Both manual and remotely actuated valves can be found in flammable liquid storage tank facilities. Fire experience has shown the importance of placement of critical shutoff valves. These valves can be either manually or remotely actuated, depending upon the type of valve and its purpose (dike drains, pipeline valve, individual storage tank valve, transfer lines, etc.).

Valves should be located so they can be operated easily and safely while remaining protected from damage (see FIGURE 3-35). Valves function best in the upright position. An inverted valve stem position eventually accumulates sediment that could cut or damage the stem. All connections should be checked periodically to ensure that they are properly connected and sealed. Secondary containment, such as collars to prevent releases from tank roof penetrations, may be used to reduce the potential for spills at valves and flanges.

Many types of hose couplings are available, especially at loading racks and product transfer points. Selection is determined by the temperature, pressure, and chemical properties of the product. Couplings must be tight to prevent spills during liquid transfer operations. Couplings designed to prevent flow when the hoses are disconnected can be used to prevent spills. Fusible link valves can be used to stop the flow on an emergency basis in case of fire. Color-coded and substance-specific couplings help prevent inadvertent connections.

## Fire Protection Systems

Flammable liquid terminals and facilities can have a range of different elements that make up the overall fire protection system (see FIGURE 3-36). Influencing factors will include the type of facility, tank design, location and contents, the overall layout of the tank farm, and fire code requirements. Fire protection measures can include the following:

- Monitoring and detection systems that alarm in the event of a spill or release
- Firewater system, including the use of deluge or water spray systems
- Foam and dry chemical systems on loading racks
- Fire proofing of critical structures to reduce failure potential (e.g., steel legs for horizontal storage tanks may be encased in concrete to insulate the steel from the heat of a fire)

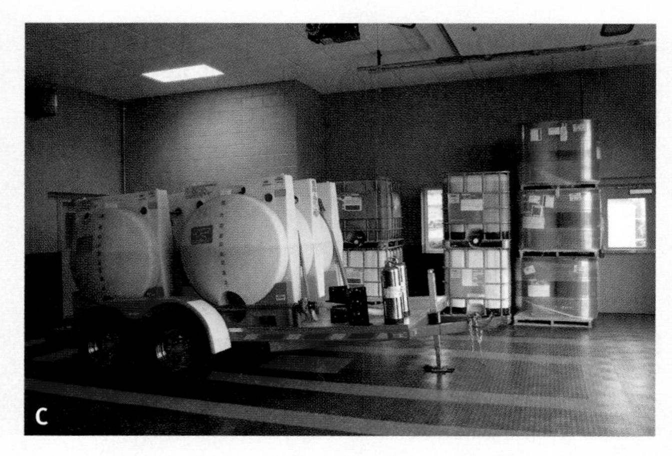

FIGURE 3-36 Fire protection features found at flammable liquid terminals and facilities can include various fixed, semifixed, and mobile equipment.

Courtesy of Gregory G. Noll.

Additional information on firefighting foam and water supply requirements can be found in Chapter 5.

## Loading Racks and Transfer Operations

Petroleum storage facilities may receive and/or ship products by pipeline, highway, rail, and marine modes of transportation. Each mode poses different tactical challenges for emergency responders.

### Cargo Tank Truck Loading Racks

Products move from storage, through a series of piping and valves, and eventually to the loading rack (see FIGURE 3-37). Product-specific dyes, detergents, and additives contained in additive tanks are added to the product at the loading rack. There are two basic types of tank truck loading rack systems used for the transfer of refined petroleum products: bottom loading and top loading.

FIGURE 3-37 Cargo tank truck loading rack operations.

Courtesy of Gregory G. Noll.

Bottom loading systems transfer product from the loading rack into the bottom of a cargo tank truck through its belly or compartment valves. It is a "closed system" and integrates a vapor recovery system so that liquid flows from the loading rack into the tank truck, while vapors flow from the tank truck back to the loading rack. The vapors are then piped to either a vapor recovery unit (VRU), where the vapors are condensed back to their liquid form, or to a vapor combustion unit (VCU), where the vapors are burnt off. Bottom loading is the most common and prevalent transfer process found today.

Top loading systems transfer product via fill stem tubes from the loading rack through the top dome covers of the tank truck. As an open system, vapors can also flow into the environment, thereby creating potential safety and environmental emissions concerns. As a result, top loading is usually limited to the transfer of lower vapor pressure products, such as diesel and fuel oils. While vapor recovery may be integrated into a top loading system, it is relatively expensive and not often found.

Safeguards that are installed on both petroleum product loading racks and MC-306/DOT-406 cargo tank trucks in order to prevent overfills include the following measures:

- Product transfer cannot begin until an electrical sensor connection is made between the loading rack pump controls and the tank truck. Although there are several manufacturers of these overfill protection systems, most people in the cargo tank industry refer to these as a "Scully System."
- At many terminals and distribution facilities, each cargo tank truck has an "electronic profile" that is stored in the loading rack software program. In addition to billing, sales, and security information, the profile includes the number of compartments, the capacity of each compartment, and the normal product(s) to be loaded.
- "Deadman" switches control product flow from the loading rack. If the deadman switch is released by the operator, the flow of product will be stopped.
- Most newer tank trucks have a bar or switch interlock connected to the vehicle braking system. When the interlock is engaged, the brakes will be engaged and the vehicle cannot be moved.
- Preset loading meters and control valves provide a method to load a specific and predetermined quantity of product.
- Product level sensors are installed in each compartment. When actuated, they will signal a high-level condition and shutoff product flow. Several types of sensing devices exist, including floats, optical sensors, and thermistors.

**FIGURE 3-38** Ethanol tank cars preparing for off-loading operations.

Courtesy of Gregory G. Noll.

Loading racks are often protected with fixed fire suppression systems, including Class B firefighting foams and dry chemical agents (e.g., potassium bicarbonate or Purple K). In addition, emergency shutoff devices (ESD) can be found at several locations in and around the loading rack.

### Railroad Loading Racks

Both crude oil and refined petroleum products are transported by rail. The largest percentage of ethanol shipments used in the formulation of various gasoline blends is moved by rail. These ethanol by rail shipments may move in blocks of tank cars or in an ethanol unit train. See **FIGURE 3-38**.

Many of the safety features found at cargo tank truck loading racks can also be found at ethanol racks where product is transferred from rail cars to storage tanks. These include the following:

- In the event of any spills or releases, product is confined in the loading rack area.
- Fixed fire detection and suppression systems. For ethanol, alcohol-resistant foam concentrates (AR-AFFF) must be used.
- Emergency shutdown devices can often be found at several locations in and around the loading rack.

### Marine Transportation and Transfer Operations

In some regions of North America, crude oil, ethanol, and refined petroleum products may be transported by tanker or barge (see **FIGURE 3-39**). There are rigid environmental protection and emergency preparedness regulations that are enforced by the U.S. Coast Guard. This includes the activation and use of Oil Spill Removal Organizations (OSRO) to manage and coordinate any spills or releases that may occur.

**FIGURE 3-39** Flammable liquids may be transferred by both tanker and barge at bulk petroleum facilities.

Courtesy of Gregory G. Noll.

## Summary

Emergency responders should understand the basic design and construction features associated with the five broad types of aboveground bulk flammable liquid storage tanks. These include the following:

- Fixed (cone and dome) roof tanks
- Open (external) floating roof tanks
- Covered (internal) floating roof tanks
- Vertical low pressure storage tanks
- Horizontal low pressure storage tanks

The type of storage tank used to store flammable and combustible liquids is determined by the physical characteristics of the product being stored and the location of the tank (e.g. tank farm vs. industrial facility).

Bulk petroleum products may be shipped via all modes of transportation, with the majority of refined petroleum products and ethanol moving via pipeline, rail, and cargo tank truck. In addition, ethanol shipments to these facilities move primarily by rail and cargo tank truck.

Causes of product releases at bulk petroleum storage facilities include valve and piping leaks, tank overfills, and natural events such as flooding and earthquakes. Leak detection for aboveground storage tank systems can be accomplished using periodic or continuous monitoring. Leak detection methods for tanks include visual inspection, inventory control, product level monitoring, subsurface monitors that monitor groundwater or soil vapor, sump monitors in dike areas, and interstitial monitors in double-wall or double-bottom tanks.

Product confinement and containment systems are designed to confine the released product to a limited area and minimize its impact upon surrounding exposures. When viewed from a systems perspective, product containment structures and options for aboveground storage tank systems include the following:

- Dike walls to minimize the spread of any spills/releases outside the immediate tank area.
- Impoundment areas where the released product flows to a remote area away from the storage tank.
- Double-walled tanks, which includes a storage tank within a larger, outer storage tank.
- Double-bottom tanks to minimize tank leakage into the ground and groundwater.
- Secondary containment, such as rubber or other impervious liners, that minimize product movement inside a dike wall into the ground and groundwater.

Ancillary equipment considerations at bulk flammable liquid facilities include piping and piping support structures, transfer pumps, valves, connections and couplings, and instrumentation. Based upon the mode(s) of transportation serving the facility, loading racks and transfer operations must be considered.

## References

1. American Petroleum Institute, *API STD 2350—Overfill Protection for Storage Tanks in Petroleum Facilities* (4th edition). Washington, DC: American Petroleum Institute (2014).
2. Emerson Process Management, *The Engineer's Guide to Overfill Protection* (2015 edition). Shakopee, MN: Emerson Process Management (2015).
3. U.S. Chemical Safety Board, "Fire Investigation Report: Caribbean Petroleum Tank Terminal Explosion and Multiple Tank Fires." Washington, DC: U.S. Chemical Safety Board (October 23, 2009).

# Incident Management and Response Considerations

CHAPTER 4

Courtesy of William T. Hand.

## Chapter Outline


- Objectives
- Key Terms
- Introduction
- Managing the Incident: The Players
- Managing the Incident: The Incident Command System
- Managing the Incident: Command Considerations
- Developing the Incident Action Plan
- Tactical Decision-Making Framework: The Eight Step Process©
- Summary
- References


## Objectives

1. Describe the general components of the National Incident Management System (NIMS) and the application of the incident command system at a petroleum storage tank emergency.
2. Describe the hazards, risks, safety procedures, and tactical guidelines for handling a storage tank fire.

## Key Terms

*Emergency Operations Center (EOC)* If the scope of an incident increases, this plant or community center may be activated. In this situation, overall coordination would be transferred to the EOC, while on-scene response operations would continue to be managed from the incident command post.

NOTE: *It is important to understand the differences and relationship between the ICP and the EOC when both are operating simultaneously at a major emergency. The ICP is primarily oriented toward tactical control issues pertaining to the on-scene response, while the EOC deals with strategic issues and coordinates logistical and resource support for on-scene operations.*

*Incident Action Plan (IAP)* The strategic goals, tactical objectives, and support requirements for the incident.

*Incident Commander (IC)* The individual responsible for establishing and managing the overall incident action plan (IAP).

*Incident Command Post (ICP)* The on-scene location where the IC develops goals and objectives, communicates with subordinates, and coordinates activities between various agencies and organizations. The ICP is the field office for on-scene response operations and requires access to communications, information, and both technical and administrative support. It may range from the front seat of an SUV to a mobile command unit. It should be identified so that arriving command and general staff personnel can easily find its location (e.g., green flag or flashing light).

*Incident Command System (ICS)* An organized system of roles, responsibilities, and standard operating procedures used to manage and direct emergency operations.

*National Incident Management System (NIMS)* The baseline incident management system established under Homeland Security Presidential Directive 5 (HSPD-5) to assist those responding to incidents or planned events. NIMS focuses upon five key areas, or components: (1) Preparedness, (2) Communications and Information Management, (3) Resource Management, (4) Command and Management, and (5) Ongoing Management and Maintenance.

*Unified Command* The process of determining overall incident strategies and tactical objectives by having all agencies, organizations, or individuals who have jurisdictional responsibility, and in some cases those who have functional responsibility at the incident, participate in the decision-making process.

## Introduction

Direct and effective command and control operations are essential at every type of incident. However, hazardous materials incidents, including bulk flammable liquid incidents, place a special burden on the command system because they often involve communications among separate agencies and the coordination of many different functions and personnel assignments—from public protective actions and the application of large volume foam streams, to environmental cleanup and recovery operations. See **FIGURE 4-1**.

This section reviews the fundamental concepts of incident management and its application at a petroleum storage tank emergency. Primary topics include the various players who characteristically appear at an incident, application of the incident command system (ICS) at a storage tank emergency, and strategic and tactical decision making. This chapter is intended to complement the information that the reader should have already gained by completing the ICS-100: Introduction to the Incident Command System and ICS-200: ICS for Single Resources and Initial Action Incidents level training courses. Where the ICS-100 and ICS-200 level courses provide the ICS "science" associated with managing hazardous materials incidents, this chapter reinforces that information by looking at the "art" of managing petroleum storage tank incidents.

**FIGURE 4-1** Establishing relationships prior to an emergency provides a solid foundation for a safe and effective emergency response.

Courtesy of Gregory G. Noll.

**FIGURE 4-2** Establishing an incident command post is a critical incident management benchmark.

Courtesy of Gregory G. Noll.

## Managing the Incident: The Players

A petroleum storage tank emergency requires different skill sets from both the emergency response community and energy industry to safely stabilize the emergency and bring it to closure. Every discipline brings their own agendas, organizational structures, and priorities to the incident scene. The seeds of a successful response start by having a preexisting relationship with your peers and other players well before you meet at the incident scene. The 0200 hours meeting at the tank farm should not be the first time you meet your response partners.

The formula for a safe and effective response is having a coordinated incident command structure in which all of the players integrate their resources to make the problem go away in a safe and effective manner. In simple terms, you need a system, and the incident command system provides the structure you need. See **FIGURE 4-2**.

The basic ICS organization that must be created to bridge these potential gaps and problems includes the following:

**Incident commander (Command or IC)** The individual responsible for establishing and managing the overall incident action plan (IAP). Depending upon the location of the storage tank emergency, the IC may come from either an industrial or public safety fire department.

**Unified commanders (UCs)** Command-level representatives from each of the primary responding agencies having authority who present their agency's interests as a member of a unified command organization. Depending on the scenario and incident timeline, they may be the lead IC or play a supporting role within the command function. The unified commanders manage their own agency's actions and make sure all efforts are coordinated through the unified command process.

For a storage tank emergency in a refinery or petrochemical process facility, on-scene command will often be unified between process operations and the emergency response team (ERT). Unified command may also be expanded depending upon the location, scope, and nature of the incident to include the lead public safety agency (e.g., fire department), the lead state/province environmental agency, and federal agencies such as the U.S. Coast Guard (USCG) and the U.S. Environmental Protection Agency (EPA).

**ICS command staff** Those individuals appointed by and directly reporting to the IC. These include the safety officer (SOFR), the liaison officer (LOFR), and the public information officer (PIO).

**ICS general staff** ICS provides a mechanism to divide and delegate tasks and develop a management structure to handle the overall control of the incident. Section chiefs are members of the IC's general staff and are responsible for the broad response functions of operations, planning, logistics, and finance/administration. Individuals below the section level are the front-line supervisors who implement tactical objectives to meet the strategies established by the IC within a branch, group, or division (e.g., hazmat group supervisor).

## The Players

Regardless of who they are and how they materialize on the scene, the IC must also be able to quickly identify and categorize the various players and participants who will interact within the ICS organization at a storage tank emergency. These can include the following:

**Facility emergency response team (ERT)** Crews of specially trained personnel used within industrial facilities for the control and mitigation of emergency situations. ERTs may be found in petroleum manufacturing and processing facilities, but are typically not encountered at flammable liquid storage and distribution facilities. When available, they will likely be the initial on-scene responders.

ERTs may consist of full-time personnel, shift personnel with ERT responsibilities as part of their job assignment (e.g., plant operators), or volunteer members. ERTs may be responsible for any combination of fire, hazmat, medical, and technical rescue emergencies, depending on the nature, size, and operation of their facility. See FIGURE 4-3.

**Fire/rescue/EMS companies** Organizations that provide resources for fire suppression, rescue, and medical triage, treatment, and transport. They implement assigned tasks, provide support to specialized assets, and help to coordinate overall response efforts. Examples include firefighters and fire officers, EMS personnel, and other knowledgeable responders on scene.

FIGURE 4-3 Petroleum manufacturing and processing facilities often have an onsite emergency response team with a robust operational capability to flow both water and foam.
Courtesy of Monroe Energy.

At a storage tank emergency, the fire service will be responsible for the delivery and movement of foam and water streams used in the control and extinguishment process. It should be noted that most public safety fire departments do not have sufficient quantities of Class B foam concentrate to sustain foam operations at a storage tank emergency, and will be reliant upon either the facility operator or emergency response contractors for the necessary foam proportioning, movement, and application resources.

**Hazardous materials response teams (HMRTs)** Teams of specially trained and medically evaluated individuals responsible for directly managing and controlling hazmat problems. They may include people from the emergency services, private industry, governmental agencies, environmental contractors, or any combination. At a storage tank emergency, they can provide risk-based assistance with health, safety, air monitoring, and sampling tasks.

**Facility managers and supervisors**—Individuals responsible for the operations and management of the bulk flammable liquid storage facility. Subordinates normally include operations and maintenance personnel, as well as other support staff. While larger facilities may be staffed for 24-hour facility operations (see FIGURE 4-4), the majority of distribution terminals and storage facilities are not staffed during "off shift" hours (i.e., evenings and weekends). Depending upon the date and time of the incident, fire department personnel may not find facility representatives on-scene upon their initial arrival.

**Emergency response contractors**—Contractors that perform a range of emergency response tasks, including incident management, health and safety management, air monitoring, provision of Class B

**FIGURE 4-4** Larger petroleum facilities, such as refineries, may have an area designated as the emergency operations center (EOC).

Courtesy of Gregory G. Noll.

foam concentrate and related foam appliances, spill control, and fire control and extinguishment. Given the limited frequency of storage tank emergencies, combined with their potential size, complexity, and resource requirements, many bulk distribution terminals and facilities have contractual agreements with emergency response contractors to assist the facility in safely controlling and extinguishing the incident. Within ICS, they may be viewed as a technical specialist.

**Environmental cleanup contractors** Organizations that provide environmental mitigation and support services at the incident. Capabilities may include ground and marine spill control capabilities, product transfer operations, site cleanup and recovery, and remediation operations. They are normally retained by either the responsible party (RP) or government environmental agencies (e.g., EPA, a state Department of Environmental Quality). Contractor personnel should be trained to meet the training requirements of OSHA 1910.120, paragraphs (b) through (o).

**Government agencies and officials** There is a high probability that any storage tank emergency will ultimately bring local, state/province, and federal government agencies with an emergency preparedness or environmental protection function to the incident. In addition to the first responder community, these can include EPA, USCG, OSHA, and DOT.

Elected officials normally do not have an emergency response function but bring a lot of political clout to the incident, especially if the scenario continues over a period of several days. Examples include mayors, city/county managers, and other elected officials who may be involved. For large-scale events they may play a coordination role, or they may delegate this responsibility to the fire chief or emergency manager. Failure to professionally address their questions and concerns within the ICS organization can have significant political and other impacts both during and after the response.

**News and social media** In today's world, virtually every person with a cell phone has the ability to communicate news and information about an incident. The collection, dissemination, and monitoring of news and information (both formal and informal) must be managed as an incident objective. Given the potential size and scope of these incidents, the establishment of a joint information system (JIS) should be strongly considered to ensure that accurate, consistent, and reliable incident information is flowing outward to information consumers and users.

**Investigators** Individuals who are responsible for determining the origin and cause of the incident, including any related evidence collection and preservation. Future legal proceedings; possible regulatory citations or criminal charges; and financial reimbursement for the time, equipment, and supplies of emergency services may well depend on investigators' efforts.

Certain types of incidents require interaction between investigators on the federal, state, and local levels, as well as in the private sector (e.g., insurance industry). Key players in the investigative process can include the local or state Fire Marshal's Office, OSHA, the U.S. Chemical Safety Board (CSB) for fixed facilities, and the National Transportation Safety Board (NTSB) if the incident involves one of the modes of transportation (e.g., marine, pipeline).

## Managing the Incident: The Incident Command System

The incident command system (ICS) is the national benchmark for the management of "all-hazard" emergency scenarios. From a regulatory perspective, OSHA 1910.120(q) requires that both public safety and industrial emergency response organizations use a "nationally recognized Incident Command System for emergencies involving hazardous materials." Another driver is Homeland Security Presidential Directive (HSPD) 5: Management of Domestic Incidents, which was released in February 2003 and establishes a single, comprehensive national incident management system. But beyond regulatory and governmental requirements, experience has shown that the normal, day-to-day business organization is not well suited to meeting the broad demands created by "working" hazmat incidents.

The National Incident Management System (NIMS) is the baseline incident management system established under HSPD-5 to assist those responding to incidents or planned events. To unite the practice of emergency management and incident response, NIMS focuses upon five key areas, or components. These individual components link together to form a larger and comprehensive incident management system. The five major components of NIMS are:

- Preparedness
- Communications and Information Management
- Resource Management
- Command and Management
- Ongoing Management and Maintenance

Discussions in this section will be limited to the ICS component of the "Command and Management" section of NIMS. ICS is an organized system of roles, responsibilities, and procedures for the command and control of emergency operations. It is a procedure-driven, all-risk system based upon the same business and organizational management principles that govern organizations on a daily basis. As is the case with the day-to-day management of any organization, ICS has both technical and political aspects that must be understood by the key players. See FIGURE 4-5.

FIGURE 4-5 Communications and information management are critical elements of NIMS. Mobile communication units (MCUs) can be a key resource in achieving communications interoperability.

Refinery Terminal Fire Company

Historically, these players were primarily concerned with the technical or operational aspects of a storage tank emergency. Today, however, the playing field has changed significantly. Storage tank emergencies can have significant political, legal, and financial effects on how both the public and the corporate shareholders view an emergency. Recent incidents have taught us that an emergency can have a favorable technical or operational outcome and still be a political disaster. In some instances, solving the operational problems may be the easiest piece of the command equation.

Scan Sheet 4-A highlights several incidents where storage tank fires resulted in tragedy.

# Scan Sheet 4-A—The Incident Command System and Site Safety Practices Save Lives

Over the years there have been several major storage tank fires that resulted in the death of multiple firefighters at a single incident. The following summarizes these tragic incidents.

**June 23, 1949**
**Perth Amboy, New Jersey**
**4 Firefighters Killed**

Heated asphalt and naphtha were stored in a dozen 2,000-gallon barrel vertical tanks without dikes or any containment. A riveted seam in a tank bottom failed 1 hour into the fire, propelling a tank 25 feet in diameter and 32 feet high into the air like a rocket. The tank fell into an adjacent tank, spreading the fire and killing members of the Emergency Response Team.

**August 18, 1959**
**Kansas City, Missouri**
**6 Firefighters Killed**

On an extremely hot day, a tank truck loading accident at a bulk plant created a spill that ignited, subsequently exposing four 21,000-gallon horizontal gasoline tanks. Three tanks contained gasoline and one kerosene, and were mounted on 9-foot-high concrete saddles. Flame impingement on the exposed tanks created vent fires, which intensified the fire. Each tank was equipped with a substandard 2-inch pressure vacuum vent. About 1-1/2 hours into the fire one of the tanks ruptured, propelling the burning tank 94 feet and spewing burning gasoline over firefighters in forward positions. Six fire firefighters were killed and over 100 people were injured. Authors Note: This historic incident was recorded on film by KMBC news cameras and was used as the basis for the widely viewed training film, "Analysis of a Bulk Plant Fire," produced by the American Petroleum Institute. Some of this dramatic footage is still used in flammable liquids training programs today.

**August 21, 1963**
**Framington, Massachusetts**
**3 Firefighters Killed**

Four horizontal storage tanks containing kerosene, fuel oil, and gasoline were involved in a bulk plant fire which started from an adjacent coal pile fire. About 30 minutes into the firefighting operation, a horizontal tank violently failed, propelling the end of the tank 200 feet. Three firefighters were killed and 20 spectators were injured including some that were over 300 feet away.

**August 17, 1975**
**Philadelphia, Pennsylvania**
**8 Firefighters Killed**

Crude oil was being off-loaded from a tanker at the Gulf Oil Refinery to a 60,000-barrel converted internal floating roof tank. The tank was filled beyond capacity, which caused a rapid emission from the tank vents. Vapors traveled to a boiler house. The vapors ignited and the fire flashed back to the crude oil outside the tank dike.

Within a brief time, an explosion occurred in the crude tank spilling additional oil into the dike. An adjoining tank containing No. 6 fuel oil became involved in the fire and several pipelines in the dike failed. During firefighting operations, water overflowed the diked area carrying crude oil onto access roads where fire apparatuses were pumping. Foam was applied onto the crude oil outside the diked area but over time, the foam blanket was disturbed by firefighters walking through the area. Crude oil floating on top of the runoff is believed to have been ignited by fire apparatuses. Eight firefighters were trapped in the burning runoff and subsequently died. Three foam trucks and two pumpers were destroyed. At least 200 members of the Philadelphia Fire Department and the Gulf Refinery Fire Brigade responded to the 11 alarms required to extinguish the fire which burned for 9 days.

For a more detailed analysis of this incident, see the following article: Smith, James P., "History as a Teacher: How a Refinery Blaze Killed 8 Firefighters." Firehouse Magazine (December 1987), pp. 16–20.

**August 31, 1976**
**Gadsden, Alabama**
**3 Firefighters Killed**

A tank truck filling accident created a running spill fire which exposed a horizontal gasoline storage tank to the fire. Approximately 40 minutes into the fire a weld seam on the storage tank failed from tank overpressure. The tank exploded and propelled the front tank head into a pumper destroying the fire apparatus. The main section of the tank traveled 240 feet, producing a fireball 600 feet in diameter and 150 feet high. Three firefighters, including the Gadsden Fire Chief, were killed by the fireball and tank fragments. Twenty-eight other people were injured.

**December 19, 1982**
**Tacoa, Venezuela**
**40 Firefighters Killed**

The storage tank was located a few miles northwest of Caracas in the small seaside village of Tacoa. The 180-foot-diameter storage tank supplied No. 6 fuel oil to an electric power generating plant. Two workers were initially killed when the tank roof blew off the 3.5-million-gallon tank.

The incident became a "spectator fire" and firefighters did not maintain effective crowd control. After 6 hours of intense burning, the tank boiled over, killing more than 150 people, of which 40 were firefighters—one of the largest loss of firefighters in history from a single incident. See the more detailed Case Study on pages 98–99.

**June 10, 1995**
**Addington, Oklahoma**
**2 Firefighters Killed**

A 55,000-barrel crude oil cone roof storage tank was struck by lightning, blowing off the roof inside the diked area. A P-19 aircraft rescue firefighting (ARFF) unit from the Sheppard AFB Texas Fire and Emergency Services Department was requested to respond via mutual aid. Approximately 9 hours after ignition, a crude oil slopover followed by the catastrophic failure of the storage tank resulted in crude oil overflowing the dike containment system. This wave of burning oil subsequently flowed downhill and entrapped two U.S. Air Force firefighters from Sheppard AFB, Texas, in their P-19 CFR vehicle. Both firefighters died from burns.

For a more detailed overview of this incident see the following:

Allen, Gene, "Two More Firefighters Die, Beware of the Crude Oil Volcano." *Industrial Fire World* (September/October 1995), p. 6.

Smith, Donald, "ATTCO Pipeline Tank Fire, Responding to the Volcanic Inferno." 1997 International Oil Spill Conference Proceedings, pp. 926–927.

## Managing the Incident: Command Considerations

Emergencies involving bulk flammable liquid storage tanks will be challenging for most public safety response agencies. Major fixed facilities, such as refineries and petrochemical plants, are likely to have an in-plant emergency response team with a robust operational capability for moving large volumes of foam and water to support response operations. In contrast, at distribution and pipeline terminals the Class B foam resources may be present, but the facility will be reliant upon either public safety agencies or emergency response contractors who specialize in bulk flammable liquid firefighting to provide the requisite operational capability.

Lessons learned from managing large, complex incidents that can be applied to petroleum storage tank emergencies include the following:

- **Command and Control.** Depending upon the size of the storage tank and the nature of the incident, storage tank emergencies will often be long and extensive operations. Effective ICs recognize that a few minutes spent establishing effective command and control at the beginning may save time and effort over the course of a long-term incident.

  Constant reassessment and possible revision of tactical operations are needed to maximize response effectiveness. The IC must plan ahead and operate with a backup plan, constantly asking questions such as these:
  - Where will the incident be in 1 hour? in 3 hours?
  - Given the current problem, what is our worst case scenario?
  - What are Plan B and Plan C if Plan A doesn't work?
- **Unified Command**. Major storage tank emergencies often involve situations where more than one organization shares management responsibility for the response, or where the incident is multi-jurisdictional in nature (see FIGURE 4-6). A unified command structure simply means that the key agencies that have jurisdictional authority or functional responsibility for an incident jointly contribute to the process of:
  - Determining a common set of incident objectives and strategies based upon incident priorities of life safety, incident stabilization, and property conservation
  - Developing a single incident action plan (IAP)
  - Maximizing use of all assigned resources
  - Resolving conflicts between the players

  If the players aren't coordinated and don't play as a team, the result is often ugly. Depending upon the location and nature of the incident, members of unified command at a storage tank emergency can include the following:
  - Local fire department(s)
  - Responsible party (RP)—senior representative from the storage tank facility
  - State Environmental Agency
  - USCG or EPA
- **Incident Potential**. The development of incident objectives and the related strategies and tactics must be based on the assessment of incident potential. Elements can include incident severity, magnitude and duration of the event, as well as the nature and degree of incident impacts. Responders will not "win the battle" if they are only focused on where the problem is at hand. They must always consider incident potential and possible growth of the problem.

  Causes of a delayed assessment of incident potential include the following:
  - Responders get so focused on the tactical problem that they fail to consider the big picture or strategic aspects of the incident.
  - Response agencies believe that a request for mutual aid or assistance might involve acknowledging a mistake.
  - Responders define the problem down to a manageable level.
  - Responders are inexperienced.
  - There is a lack of information to make an informed decision or information is not shared across the command structure or between responding agencies.
- **Decision Making.** The decision-making process begins with the IC and operations section chief recognizing the need to avoid dead-end decisions. Whenever possible, decisions must be open ended, allowing for expansion or reversal. Recognize the following:
  - *Distinguish between assumptions and facts*. Response operations must sometimes be based upon incomplete or assumed information. Factual information is often not available or incomplete, particularly in the initial stages of an emergency. Ask these basic questions: What do we know (i.e., confirmed facts)? What do we think we know that must be confirmed (i.e., assumptions)? What don't we know that we need to determine (i.e., gaps)?

FIGURE 4-6 Establishing a unified command organization between the facility and public safety response agencies is a key incident management benchmark.

Courtesy of Michael S. Hildebrand.

- *Maintain a flexible approach to decision making.* The overall IAP must be constantly updated as more and better information is received from operations units and outside sources. Feedback allows for revisions to the general strategy, specific tactics, and all major decisions.
- *Shift to a management role after initiating action.* An IC cannot make all ongoing response decisions. The efficiency of command decisions will improve once the IC delegates tactical responsibility. Otherwise Command will quickly become overwhelmed with both people and information.

The IC must quickly prioritize problems and develop solutions. This requires the effective gathering, recording, and organizing of information.

- **Technical Specialists**. For public safety responders, the likely source of hazard information at facilities without an onsite ERT will be technical specialists from the involved facility (operations supervisor, product or container specialists, etc.). These individuals usually understand both the product and the container, the upstream and downstream impacts of the incident, and operational considerations that must be considered in "making the problem go away."
- **Long-Term Incidents and Planning.** Storage tank emergencies that extend over multiple operational periods will create different challenges for emergency responders. Issues such as developing a shift schedule, determining short-term and long-term logistical requirements, and establishing a formal IAP development flow and process are foreign to many responders.

Where can you go for help? This is the focus of the ICS 300: Intermediate ICS for Expanding Incidents for Operational First Responders course. In addition, a number of regions and states have established concepts such as incident support teams (ISTs) and incident management teams (IMTs) to assist local ICs with the management of complex or long-term emergencies.

## Developing the Incident Action Plan

This section provides the basic strategic and tactical framework for decision making at storage tank emergencies. It is based on material covered in Chapters 5 through 12 of *Hazardous Materials: Managing the Incident* (4th edition). If this is your first time digging into this topic, we recommend that you familiarize yourself with the material in these chapters for a more detailed explanation.

The incident action plan is developed based upon the IC's assessment of (1) incident potential (i.e., visualizing hazardous materials behavior and estimating the outcome of that behavior), and (2) the initial operational strategy.

FIGURE 4-7 The ability to move and apply large volumes of water will be a key element in developing the operational strategy.

Courtesy of Monroe Energy.

Every incident must have an oral or written IAP. The IAP should clearly state the strategic goals, tactical objectives, and assignments that must be implemented to control the problem, as well as required resources and support materials. The IAP provides all command and supervisory personnel with the direction of future actions. See FIGURE 4-7.

**Strategic goals** are the broad game plan developed to meet the incident priorities (life safety, incident stabilization, environmental and property conservation). Essentially, strategic goals are, "What are you going to do to make the problem go away?"

Several strategic goals may be pursued simultaneously during an incident. Examples of common strategic goals at a storage tank emergency include the following:

- Rescue
- Public Protective Actions
- Spill Control (Confinement)
- Leak Control (Containment)
- Fire Control
- Recovery

**Tactical objectives** are specific and measurable processes implemented to achieve the strategic goals. In simple terms, tactical objectives are the "How are you going to do it?" side of the equation. Tactical objectives are then eventually tied to specific tasks that are assigned to particular response units, such as fire, rescue, law enforcement, etc. For example, tactical objectives for a spill control strategy would include diking, damming, and retention.

Tactical response objectives to control and mitigate the problem may be implemented in either an offensive, defensive, or nonintervention mode. Criteria for evaluating these options include the level of available resources (e.g., personnel and equipment), the level of training and capabilities of emergency responders, and the potential harm created by the problem.

- **Offensive Mode**. These are aggressive leak, spill, and fire control tactics designed to quickly control or mitigate the emergency. Although increasing the risks to responders, offensive tactics may be justified if rescue operations can be quickly achieved, if the spill can be rapidly contained or confined, or the fire quickly extinguished. The success of an offensive mode operation is dependent on having the necessary resources available in a timely manner. In a storage tank emergency, this often relates to the ability of responders to initiate and sustain large-volume Class B foam and water operations.
- **Defensive Mode**. These are less aggressive spill and fire control tactics where certain areas may be conceded to the emergency, with response efforts directed toward limiting the overall size or spread of the problem. Defensive tactics decrease the risk to responders and may be employed as methods of either reducing the size of the spill or reducing the pressure of the affected pipeline or tank. Examples include isolating a pipeline by closing remote valves, shutting down pumps, and exposure protection.
- **Nonintervention Mode**. Nonintervention is essentially "no action." Essentially, the risks of intervening are unacceptable when compared to the risks of allowing the incident to follow a natural outcome, such as scenarios with a boilover or explosion potential. In other cases, environmental impacts may be reduced by allowing a flammable liquids fire to burn itself out. All personnel are withdrawn to a safe location, with unmanned master streams left in place to protect exposures.

Storage tank fires can be dynamic events. If response operations are not achieving their desired results during the fire, the IC should be prepared to consider changing the strategy and tactics to safely achieve operational success. See Figure 4-7 for strategies and tactics for storage tank response operations. **TABLE 4-1** Strategies and tactics for storage tank response operations.

In some situations, nonintervention tactics may be implemented until sufficient resources arrive on-scene and an offensive attack can be implemented. Most operations at a storage tank emergency will begin from a defensive point of view. The most important question the IC should ask is, "What happens if I do nothing?" There will also be times when an operation is in a marginal mode. In other words, initial information indicates that it is relatively safe to attempt an offensive tactical objective, yet it is very possible that things may turn for the worse during that process.

## Size-Up and Logistical Considerations

Storage tank incidents require an effective size-up and strong logistical support throughout the incident. Major tank fires require large numbers of personnel, fire apparatuses, water supplies, and foam concentrate to mitigate the incident. Successful tank firefighting operations require the right amount of foam concentrate, proportioned at the correct rate, with the right type of foam, for the prescribed duration or the fire will not go out.

Storage tank fires can be campaign events that can potentially last hours or even days. Like any large fire or long-term incident, response operations must be supported with a logistics plan that can sustain firefighters (e.g., rehab, food, and crew rest) and fire apparatuses (fuel and mechanics) for the duration of the incident. Establishment of a logistics section early in the incident will be critical in identifying and meeting these challenges well before they are needed.

This section will provide an overview of the key items to consider when conducting the initial size-up of a bulk storage tank incident. The major emphasis in this section will be on the logistical issues associated with large bulk storage tank fires as opposed to smaller fire incidents or major non-fire oil spill type incidents, although many of the considerations are the same. We will also take an in-depth look at the requirements for water supply and foam concentrate to extinguish large-scale fires in Section 5. Specific tactical considerations for various types of storage tank incidents will be discussed in Section 6.

## Initial Planning and Size-Up

### Preincident Plans

The initial size-up of a large storage tank fire should begin long before the incident through a structured preplanning process. Preincident plans (or preplans) are individually prepared documents that focus on a specific tank in a specific facility. Preplans are significantly different from emergency response plans, because they provide specific information for the IC concerning the problems associated with a specific tank and they outline detailed information concerning the tactics, fire flows, and foam concentrate requirements needed to extinguish the fire. In contrast, the emergency response plan defines the general emergency response procedures for the entire facility. Using a football analogy, the preincident plan is the quarterback's playbook used on gameday, whereas the emergency response plan is the league's rulebook. The emergency response plan defines how the game is played, who the players are, who makes the key decisions, what type of equipment is used, and so on.

**TABLE 4-1** Srategies and tactics for storage tank response operations.

| ORGANIZATIONAL LEVELS | STRATEGIC GOALS | TACTICAL OBJECTIVES | OPERATIONAL TASKS |
|---|---|---|---|
| | "What are you going to do?" | "How are you going to do it?" | "Do it." |
| DEFINITION | STRATEGY: The overall plan to control the incident and meet incident priorities. | TACTICS: The specific and measurable processes implemented to achieve the strategic goals. | TASKS: The specific activities that accomplish a tactical objective. |
| KEY ELEMENTS | ■ Goals<br>■ Overall game plan<br>■ Broad in nature<br>■ Meets incident priorities (life safety, incident stabilization, environmental and property conservation) | ■ Objective-oriented<br>■ Specific and measurable<br>■ Often builds upon procedures | ■ "Hands-on" work to meet the tactical objectives.<br>■ The most important organizational level—where the work is actually performed.<br>■ Most problems "go away" as a result of members performing task-level activities. |
| DECISION MAKERS | ■ Incident Commander<br>■ Unified Command<br>■ Section Chiefs | ■ Operations Section Chief<br>■ Group/Division Supervisors | ■ Individual units and individuals |
| OPTIONS | ■ Rescue<br>■ Public Protective Options<br>■ Fire Control<br>■ Spill Control<br>■ Leak Control<br>■ Recovery | ■ Rescue<br>■ Public Protective Options<br>• Evacuation<br>• Protection-in-Place<br>■ Fire Control<br>• Exposure Protection<br>• Extinguishment<br>■ Leak Control<br>• Patching and Plugging<br>• Pressure Isolation and Reduction | ■ Spill Control (Confinement)<br>• Diking, Damming, & Diversion<br>• Retention<br>• Vapor Suppression |

Planning efforts should be practical and problem areas prioritized using a hazards analysis process. Preincident plans for storage tank facilities should be developed based on the following:

- *Type of hazards and risks present*. This should include the size and diameter of the tanks, products being stored, product transfer modes used for moving product into and out of the storage facility, and fixed or semifixed fire protection systems.
- *Exposures*. This would include critical infrastructure, sensitive receptors, and environmental vulnerabilities. Sensitive receptors could include schools, nursing homes, and hospitals, while critical infrastructure could include adjacent transportation and energy distribution corridors. Environmental vulnerabilities include waterways, wetlands, and sole source aquifers.
- *Locations with poor water supplies*. Offensive and defensive strategies will have a low probability of success in the absence of large and sustained water supplies. The assessment of water supply capabilities should include amount of water available, size and type of water pumps including their energy sources, size of distribution mains, and the ability of emergency responders to move large volume flows. In addition, the ability to initiate dewatering operations to control runoff from firefighting operations must also be considered.
- *Tanks requiring large quantities of foam concentrate such as bulk petroleum storage facilities, pipelines, etc*. The assessment of Class B foam capabilities should include type and quantity of foam concentrate required based upon the product and size of the tank/hazard, foam appliances needed and available, foam application devices available, foam logistics operations, and the ability to sustain foam application operations for at least 1 hour.
- *Response routes and storage tank accessibility*. Emergency responders should evaluate both response route(s) into a bulk storage facility and the ability to get vehicle and equipment access in proximity to the storage tank hazard to initiate firefighting operations. Response constraints can include single approach and access corridors, overhead cable trays and piping systems, railroad tracks, draw bridges, etc. Tank accessibility constraints include the size, width and weight capacity of access roads, and ground slope and earthen areas that cannot support the weight of either fire apparatuses or the placement of aerial unit outriggers.

Scan Sheet 4-B provides an example of a preincident plan developed for a petroleum distribution terminal.

# Scan Sheet 4-B — Preplans Should Provide Specific Tactical Information

Courtesy of Gulf Oil Limited Partnership

FACILITY FIRE ACTION PLAN: HARRISBURG TERMINAL CONTROL NUMBER: HG001 REVISION: 0 (09/29/2006)

FACILITY STORAGE TANK INFORMATION

| TANK NUMBER | PRODUCT | DIKE LOCATION | CAPACITY (GALLONS) | TANK SIZE | | SURFACE AREA | FLOATING ROOF | | FLOATING ROOF TYPE | FOAM CHAMBERS | | APPLICATION RATE | REQUIRED FLOW (GPM) | CONCENTRATE (GPM 1%) | TOTAL GALLONS CONCENTRATE (65 MIN 1%) | CONCENTRATE (GPM 3%) | TOTAL GALLONS CONCENTRATE (65 MIN 3%) |
|---|---|---|---|---|---|---|---|---|---|---|---|---|---|---|---|---|---|
| | | | | DIAMETER | HEIGHT | | YES | NO | | YES | NO | | | | | | |
| 1 | GASOLINE | 2 | 699,636 | 58'-0" | 48'-0" | 1,963' | X | | ALUMINUM | | X | 0.16 | 350 | 3.5 | 228 | 10.5 | 683 |
| 2 | GASOLINE | 2 | 298,578 | 38'-0" | 42'-0" | 962' | X | | ALUMINUM | | X | 0.16 | 175 | 1.75 | 114 | 5.25 | 342 |
| 3 | TRANS MIX | 1 | 396,852 | 36'-0" | 42'-0" | 1,017' | X | | ALUMINUM | | X | 0.16 | 185 | 1.85 | 120 | 5.55 | 361 |
| 4 | KEROSENE | 1 | 213,990 | 30'-0" | 42'-0" | 707' | | X | NA | | X | 0.16 | 150 | 1.5 | 98 | 4.5 | 293 |
| 5 | KEROSENE | 1 | 101,556 | 24'-0" | 30'-0" | 452' | | X | NA | | X | 0.16 | 115 | 1.15 | 75 | 3.45 | 225 |
| 6 | GASOLINE | 3 | 1,721,244 | 80'-0" | 48'-0" | 5,024' | X | | ALUMINUM | | X | 0.16 | 800 | 8 | 520 | 24 | 1,560 |
| 7 | HEAT OIL | 3 | 3,085,866 | 98'-0" | 56'-0" | 7,539' | | X | NA | | X | 0.16 | 1,300 | 13 | 845 | 39 | 2,535 |
| 8 | WATER | 2 | 20,000 | NA | NA | NA | NA | NA | NA | | X | NA | NA | NA | NA | NA | NA |
| 9 | KNOCK-OUT | 2 | NA | NA | NA | NA | NA | NA | NA | | X | NA | NA | NA | NA | NA | NA |
| 19 | GASOLINE | 3 | 654,784 | 65'-0" | 36'-0" | 3,317' | X | | ALUMINUM | | X | 0.16 | 575 | 5.75 | 375 | 17.25 | 1,122 |
| 22 | DIESEL | 3 | 1,014,594 | 60'-0" | 48'-0" | 2,826' | | X | NA | | X | 0.16 | 500 | 5 | 325 | 15 | 975 |

NOTES

This chart shows foam required for a one time full surface extinguishment only. Additional foam will be required for post extinguishment and re-supply.

NO FOAM CHAMBERS on any of the storage tanks. Foam calculations using chart from Williams Fire & Hazard Control.

NA: Not applicable.

Columns filled in light green contain 1% foam concentrations only for tank fires.

Columns filled in light blue contain 3% foam concentrations only for tank fires.

GASOLINE — Tank cells filled in red contain gasoline only.

DIESEL/ HEAT OIL — Tank cells filled in yellow can contain diesel or heating oil.

KEROSENE — Tank cells filled in magenta contain kerosene only.

TRANS MIX — Tank cells filled in orange contain trans mix (gasoline and diesel mix) only.

Courtesy of Gulf Oil Limited Partnership

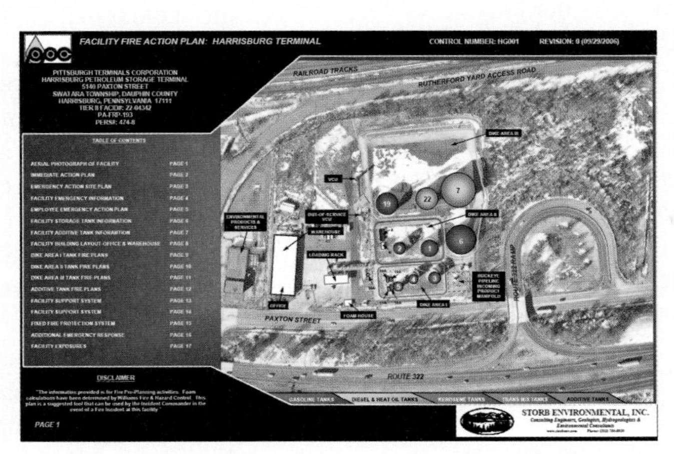


Courtesy of Gulf Oil Limited Partnership

### Initial Size-Up

The initial response and the size-up of a storage tank emergency will be a challenging experience for most emergency responders. Smoke, noise, bad weather, darkness, and stress present certain challenges for acquiring and processing information quickly and effectively. Similarly, experience has shown that some of the information provided by people at the scene of the incident when the problem started is often not accurate. It is important that a structured process be used to size up the problem before committing resources.

Emergency scenarios involving aboveground petroleum storage tanks can fall into two broad scenarios:

1. *Non-fire scenarios*, where the storage tank has experienced some type of container failure or release which can result in either environmental impacts (no ignition of the spill) or the increased risk of a post-release ignition scenario.

   The risks of ignition will vary depending upon the product involved, the quantities released, product containment and proximity to exposures, and ignition sources. Examples of non-fire scenarios can include a partially sunken roof on an open floating roof tank, a tank overfill with product controlled within the containment area, and leaks or spills from the storage tank or its related auxiliary equipment and piping that may be contained inside the dike area or escape outside the dike area.
2. *Fire scenarios* are where ignition has already taken place prior to the arrival of emergency responders. Examples of fire scenarios can include the following:
   - Tank overfill fire that involves a combination of the storage tank, dike, and ground areas
   - Vent fire on a fixed roof tank
   - Seal fire on an open floating roof tank as a result of a lightning strike
   - Full liquid surface fire that is either obstructed or unobstructed by the position of the roof

Critical information that should be acquired by the IC during the size-up process should include the following:

- The time the incident started (This may not necessarily be the same time the incident was reported!)
- The time emergency responders arrived on-scene
- Probability that the fire will be confined to its present size and how fast the fire will progress over the next 10, 20, 30, and 60 minutes
- Product(s) involved (flammable or combustible liquid, polar solvent, or chemical), including the quantity, surface area involved, and depth of the product/spill
- Physical and chemical properties of the materials involved, including flash point, boiling point, solubility (e.g., hydrocarbon or polar solvent), and specific gravity
- Estimated preburn time (i.e., how long the fire has been burning), to help the IC determine factors such as how "hot" the fuel and tank shell are, identify and prioritize exposures, consider transfer and pump off options, determine if a heat wave is developing for crude oils, etc.
- Layout of the incident, including the following specific points:
  - Type of storage tank(s) involved (See Section 3 for storage design and construction considerations.)
  - Size and capacity of the containment area(s) surrounding the involved storage tank(s), including the valve status of any drain valves (open or closed)
  - Valves, piping systems, and auxiliary equipment involved or exposed by the incident
  - Surrounding exposures, including tanks, buildings, process units, utilities, etc., which should include identifying and prioritizing exposures (e.g., flame impingement vs. radiant heat exposure), of which facility personnel can be of assistance in making these decisions

## Tactical Decision-Making Framework: The Eight Step Process©

On-scene response operations should be based on a structured and standardized system of protocols and procedures. Regardless of the nature of the incident and response, a reliance on standardized procedures will bring consistency to the tactical operation and will help to minimize the risk of exposure and harm to all responders.

The Eight Step Process© outlines the basic tactical functions to be evaluated and implemented at incidents involving hazardous materials. Like all SOPs, the Eight Step Process© should be viewed as a flexible guideline and not as a rigid rule. Individual departments and agencies should decide what works best for them.

The Eight Step Process© offers several benefits. First, it recognizes that the majority of incidents involving hazardous materials are minor in nature and generally involve limited quantities. It also builds on the action of first responding units and facilitates identifying the roles and responsibilities of each level of response. The Eight Step Process© provides a flexible management system that expands as the scope and magnitude of the incident grows and, finally, it provides a consistent management structure, regardless of the classes of hazardous materials involved.

Essentially, there are eight basic functions that must be evaluated at hazardous materials emergencies, including storage tank emergencies. These eight functions typically follow an implementation timeline at the incident:

1. Site Management and Control
2. Identify the Problem
3. Hazard Assessment and Risk Evaluation
4. Select the Personal Protective Clothing and Equipment
5. Information Management and Resource Coordination
6. Implement Response Objectives
7. Decontamination and Cleanup Operations
8. Terminate the Incident

A model tactical procedure for storage tank emergencies based on the Eight Step Process© is provided as a general guideline for preparing a local standard operating procedure.

## Step 1: Site Management and Control

**FUNCTION**: Site management and control involves managing and securing the physical layout of the incident. The operational reality is that you cannot safely and effectively manage the incident if you do not have control of the scene. As a result, site management and control is a critical benchmark in the overall success of the response and is the foundation on which all subsequent response functions and tactics are built. See FIGURE 4-8.

**GOAL**: Establish the playing field so that all subsequent response operations can be implemented both safety and effectively.

FIGURE 4-8 Step 1 involves managing and securing the physical layout of the incident, including access control.

Courtesy of Gregory G. Noll.

### CHECKLIST

- During the approach to the incident scene, avoid committing or positioning personnel and units in a hazardous position. Assess the situation and consider having an escape route out of the area if conditions should deteriorate suddenly.
- Upon arrival:
  - Determine the need for rescue.
  - Determine the extent of the hazard area.
  - Establish a security perimeter.
  - Designate hazard control zones.
  - Assume command and control of the incident, including establishment of an incident command post (ICP).
- Establish a staging area (Levels I, II) for additional responding equipment and personnel.
- Establish an isolation perimeter (i.e., outer perimeter) to isolate the area and deny entry. Establish access control—restrict emergency site access to authorized essential personnel; all nonessential personnel should be isolated from the problem. Isolation perimeters should include land, water, and air areas.
- Establish a hot zone or inner perimeter as the "playing field." The location of the inner hot zone should be identified and communicated to all personnel operating on the site.
- Do not attempt to enter the area unless you have the appropriate level of respiratory and skin protection, based on the hazards present. If people are down, self-contained breathing apparatuses (SCBAs) will be considered the minimum level of respiratory protection for initial emergency response operations.
- Initiate public protective actions (i.e., evacuation, protection-in-place), as appropriate.

### *RESPONDER TIPS*

- Site management establishes the playing field for the players (responders) and the spectators (everyone else).
- The initial 10 minutes of the incident will determine operations for the next 60 minutes, and the first 60 minutes will determine operations for the first 8 hours.
- Large releases, such as flammable liquid spills, may rapidly leave diked areas and migrate out of the initial spill area by way of open dike drains, sewers, and roadways.
- Special attention should be given to the possibility that spilled materials may migrate near or under operating emergency vehicles and equipment. During approach to the incident scene, avoid committing or

positioning personnel and vehicles in a hazardous position or situation.
- Do not permit emergency responders to operate from inside diked areas or from locations where flammable liquids have accumulated. Ignition of flammable vapors may trap personnel performing rescue or firefighting duties.
- Designate an emergency evacuation signal and identify rally points if emergency evacuation is necessary.

## Step 2: Identify the Problem

FUNCTION: Identify, confirm, and verify the scope and nature of the problem. This includes the type of storage tank(s) and product involved, and surrounding exposures. See FIGURE 4-9.

GOAL: Identify the scope and nature of the problem, including the type and nature of hazardous materials involved as appropriate. If more than one tank or piping system is involved, each individual component should be identified.

CHECKLIST

- Survey the incident—identify the nature and severity of the immediate problem, including the recognition, identification, and verification of the material(s) involved, type of container involved, and any potential or existing life hazards. If multiple problems exist, prioritize them and make independent assignments.

FIGURE 4-9 The goal of Step 2 is to identify the scope and nature of the problem, including the type of storage tank and product(s) involved.

Courtesy of Jorge Carrasco.

- Operators, supervisors, and witnesses should be identified and interviewed to obtain as much information as possible.

*RESPONDER TIPS*

- A problem well defined is half solved.
- Assume that initial information is not correct. Always verify your initial information. Verify, verify, verify.

## Step 3: Hazard Assessment and Risk Evaluation

FUNCTION: This is the most critical function that public safety personnel perform (see FIGURE 4-10). The primary objective of the risk evaluation process is to determine whether or not responders should intervene, and what strategic objectives and tactical options should be pursued to control the problem at hand. You can't get this wrong. If you lack the expertise to do this function adequately, get help from someone who can provide that assistance, such as local HMRTs and product/container technical specialists.

GOAL: Assess the hazards present, evaluate the level of risk, and establish an IAP to make the problem go away.

CHECKLIST

- Assess the hazards posed by the problem (health, physical, chemical, other). Most hydrocarbon-based flammable liquids have specific gravities less than 1, and will float on the surface of water. In contrast, ethanol is a polar solvent and is water miscible.
- Environmental conditions, including runoff, wind, precipitation, and topography should be monitored. Will the runoff be hazardous to responders?

FIGURE 4-10 Step 3, assessing the hazards and risks, is the most critical function that emergency responders perform.

U.S. Air Force

- Conduct air monitoring to determine the concentration of contaminants, their location, and the identification of safe, unsafe, and dangerous areas.
- Evaluate the risks associated with the incident. Factors that may influence the level of risks include:
  - Presence of potential ignition sources
  - Amount of product released and total surface area involved
  - Type of storage tank
  - Probability that the fire/release will be confined to its present size
  - Proximity of exposures
  - Current and forecasted weather conditions (e.g., high winds, wind direction, lightning storm)
  - Level of available resources (e.g., water, foam concentrate) to mitigate the problem safely and their approximate response time
- Based on the risk evaluation process, develop your IAP. Determine whether the incident should be handled using offensive, defensive, or nonintervention strategies. Remember that offensive tactics increase the risks to emergency responders.
  - *Offensive Mode.* Require responders to control/mitigate the emergency from within/inside the area of high risk.
  - *Defensive Mode.* Permit responders to control/mitigate the emergency remote from the area of highest risk.
  - *Nonintervention Mode.* Pursue a passive attack posture until the arrival of additional personnel or equipment, or allow the fire to burn itself out.

| STRATEGY | OFFENSIVE | DEFENSIVE | NONINTER-VENTION |
|---|---|---|---|
| Rescue | X | | |
| Public Protective Actions | X | X | X |
| Spill Control | X | X | |
| Leak Control | X | | |
| Fire Control | X | X | |
| Cleanup and Recovery | X | X | |

*RESPONDER TIPS*

- Focus on those things that you can change and that will make a positive difference to the outcome.
- Every incident will arrive at some outcome, with or without your help. If you can't change the outcome, why get involved? Doing nothing and letting the incident take its course without intervention may be a viable option.
- There's nothing wrong with taking a risk. If there is much to be gained, there is much to be risked. If there is little to be gained, then little should be risked.
- Hour 1 priorities within the IAP are as follows:
  - Establish site management and control.
  - Determine the type of storage tank and product(s) involved.
  - Ensure the safety of all personnel from all hazards.
  - Ensure that PPE is appropriate for the hazards.
  - Initiate tactical objectives to accomplish initial rescue and public protective action needs.
- Remember the PACE model of planning—Primary plan, Alternate plan, Contingency plan(s), and Emergency plan.

## Step 4: Select the Personal Protective Clothing and Equipment

**FUNCTION**: Based on the results of the hazard and risk assessment process, emergency response personnel will select the proper level of personal protective clothing and equipment. See FIGURE 4-11.

**GOAL**: Ensure that all emergency response personnel have the appropriate level of personal protective clothing and equipment (skin and respiratory protection) for the expected tasks to be performed.

CHECKLIST

- The selection of personal protective clothing will depend on the hazards and properties of the materials involved and the response objectives to be implemented (i.e., offensive, defensive, or nonintervention).

FIGURE 4-11 The goal of Step 4 is to ensure that all emergency response personnel have the appropriate level of PPE for their expected tasks.

Refinery Terminal Fire Company

- For storage tank emergencies involving fire and working in the established hot zone, structural firefighting gear and SCBA will be required. This includes helmet, fire retardant hood, turnout coat and pants, personal alert safety system (PASS device), and gloves. Positive-pressure SCBA should be considered the minimum level of respiratory protective clothing. Support activities outside the hot zone may allow for a lower degree of PPE.
- Do not place personnel in an unsafe emergency condition.

*RESPONDER TIPS*

- Structural firefighting protective clothing is not designed to provide protection against chemical hazards.
- Wearing any type and level of impermeable protective clothing creates the potential for heat stress injuries.
- Personal protective clothing is NOT your first line of defense. . . it is your last line of defense!

## Step 5: Information Management and Resource Coordination

**FUNCTION:** Refers to proper management, coordination, and dissemination of all pertinent data and information within the ICS in effect at the scene. In simple terms, this function cannot be effectively accomplished unless a unified ICS organization is in place. Of particular importance is the ability to determine the incident factors involved, which functions of the Eight Step Process© have been completed, what additional information must be obtained, and what incident factors remain unknown. See FIGURE 4-12.

FIGURE 4-12 Information management and resource coordination cannot occur in the absence of an effective on-scene ICS organization.

Courtesy of Gregory G. Noll.

**GOAL:** Provide for the timely and effective management, coordination, and dissemination of all pertinent data, information, and resources between all of the players.

**CHECKLIST**

- Confirm that the ICP is in a safe area. Personnel not directly involved in the overall command and control of the incident should be removed from the ICP area.
- Confirm that a unified command organization is in place and all key response and support agencies are represented directly or via a liaison officer.
- Ensure that emergency response plans are coordinated with facility process/tank farm operations.
- Expand the ICS and create additional branches, divisions, or groups, as necessary.
- Ensure that the appropriate internal and external notifications have been made.
- Coordinate information and provide briefings to other agencies, as appropriate.
- Confirm emergency orders and follow through to ensure that they are fully understood and correctly implemented. Maintain strict control of the situation.
- Make sure that there is continuing progress toward solving the emergency in a timely manner. Do not delay in calling for additional assistance if conditions appear to be deteriorating.
- If activated, provide regular updates to the local/facility emergency operations center (EOC).

*RESPONDER TIPS*

- Don't look stupid because you didn't have a plan.
- Bad news doesn't get better with time. If there's a problem, the earlier you know about it, the sooner you can start to fix it.
- Don't allow external resources to "freelance" or do the "end run."
- Don't let your lack of a planning section become the Achilles heel of your response. Establish it early, particularly if the incident has the potential to become a "campaign event."
- Operations may win battles, but logistics wins wars. Incidents that have specialized or substantial resource needs require the logistics section to be activated as early as possible.

## Step 6: Implement Response Objectives

**FUNCTION**: In this phase, responders implement the best available strategic goals and tactical objectives that will produce the most favorable outcome. If the incident is in the emergency phase, this is where we "make the problem go away." Common strategies to protect people and stabilize the problem include rescue, public protective

FIGURE 4-13 The goal of Step 6 is to ensure that the incident priorities and IAP are carried out in a safe, timely, and effective manner.

Courtesy of Gregory G. Noll.

actions, spill control, leak control, fire control, and recovery operations. See FIGURE 4-13.

If the incident is in the post-emergency response phase, the focus of response personnel will likely become scene safety, cleanup, evidence preservation (as appropriate), and incident investigation. Specific tasks will include (1) initial site entry and monitoring to determine the extent of the hazards present, (2) an evaluation of the scene to locate evidence that can be used to reconstruct the events leading up to the incident, (3) identification of the contributing factors that caused the incident, (4) interviewing of on-scene personnel and witnesses to corroborate the information obtained and opinions formed based on the available data, and (5) documentation of preliminary results.

GOAL: Ensure that the incident priorities (i.e., rescue, incident stabilization, environmental and property protection) are accomplished in a safe, timely, and effective manner.

### CHECKLIST

- Implement response objectives. Remember that offensive tactics increase the risks to emergency responders; evaluate the risks of offensive control tactics before sending emergency response crews into the hazard area.
  - *Offensive Mode.* Require responders to control/mitigate the emergency from within/inside the area of high risk.
  - *Defensive Mode.* Permit responders to control/mitigate the emergency remote from the area of highest risk.
  - *Nonintervention Mode.* Pursuing a passive attack posture until the arrival of additional personnel or equipment, or allowing the fire to burn itself out.

| STRATEGY | OFFENSIVE | DEFENSIVE | NONINTERVENTION |
|---|---|---|---|
| Rescue | X | | |
| Public Protective Actions | X | X | X |
| Spill Control | X | X | |
| Leak Control | X | | |
| Fire Control | X | X | |
| Cleanup and Recovery | X | X | |

NOTE: Rapidly changing incident conditions may require using multiple tactics simultaneously or switching from one tactic to another. Defensive tactics are always desirable over offensive tactics if they can accomplish the same objectives.

#### Exposure Protection Considerations

- Protect exposures at the emergency scene. Anticipate accidental spill ignition of flammable liquids. Exposures should be evaluated and prioritized so that water supplies and emergency response teams are conserved and used correctly.
- Exposures should be prioritized in the following manner:
  - *Primary Exposures:* Vessels, piping systems, or critical support structures exposed to direct flame impingement. Failure of adjacent vessels, tanks, and piping systems is likely unless cooling water is applied.
  - *Secondary Exposures:* Vessels, piping systems, or critical support structures exposed to radiant heat. Failure of structural components is possible if cooling water is not applied.
  - *Tertiary Exposures:* Noncritical equipment without life safety implications.
- Once spill exposures have been prioritized, the IC should develop a water and foam concentrate supply plan as required, in conjunction with facility utility and maintenance personnel. The water supply group should ensure that fire pumps are operating and that an uninterrupted water and foam supply can be delivered for the duration of the emergency.
- Water and foam concentrate supplies must be maintained for significant periods after the spill has been extinguished to ensure that ignition of the flammable liquid spill does not occur.
- Water required for protecting exposures and vapor suppression may create significant water runoff problems within the facility. The IC should anticipate local flooding conditions and evaluate the possibility of runoff becoming contaminated with free-floating hydrocarbons. Application of

firefighting foam to runoff may be required to prevent ignition.
- Conduct regular air monitoring of the hazard area to determine if conditions are changing.

### Firefighting Considerations

- Hydrocarbon spill fires will require foam application rates ranging from 0.10 to 0.16 gpm/sq ft depending on the type of foam concentrate used according to NFPA guidelines. Polar solvents will require at least 0.20 gpm/sq ft. Based upon the diameter of the storage tank, even greater application rates may be required. The foam manufacturer should be consulted for proper application rates and other considerations.
- As a general tactical guideline, foam application for vapor suppression should begin immediately even if enough foam concentrate is not onsite to cover 100% of the exposed flammable liquid surface area. Partial coverage of the spill will have an immediate effect on the production of flammable vapors and may prevent accidental ignition.
- Under special circumstances, the facility personnel may not be able to remotely isolate the source of the flammable liquid. It may be necessary for the emergency responders to manually isolate the sources of flammable liquid under the protection of hoselines. Before entry into the hot zone is permitted, responders must be properly protected: use the buddy system, have a plan for escape, have backup personnel in place, and operate with the knowledge of the IC.

  Foam must be applied prior to and during entry, to decrease danger to the entry team. Foam blankets should be aspirated to provide a longer lasting foam blanket. However, combination nozzles may be a preferred option for entry operations to isolate or close a valve, as the hoseline can be adjusted to control the fuel at its source based upon changing incident conditions.
- Foam blankets applied to flammable liquid spills for vapor suppression will deteriorate over time and must be continuously maintained. Water used for cooling exposures can destroy foam blankets.
- Air monitoring should continue until the emergency has been terminated by the on-scene IC. Additional monitoring may be required as the site is being restored to normal operation.

*RESPONDER TIPS*

- What will happen if I do nothing? Remember—this is the baseline for hazmat decision making and should be the element against which all strategies and tactics are compared.
- Surprises are nice on your birthday but not on the emergency scene. Always have a Plan B in case Plan A doesn't work!

## Step 7: Decontamination and Cleanup Operations

**FUNCTION**: Decontamination (decon) is the process of making personnel, equipment, and supplies "safe" by reducing or eliminating harmful substances (i.e., contaminants) that are present when entering and working in contaminated areas (i.e., hot zone or inner perimeter). Although decon is commonly addressed in terms of "cleaning" personnel and equipment after entry operations, in some instances, due to the nature of the materials involved, decontamination of clothing and equipment may not be possible and these items may require disposal.

All personnel trained to the First Responder Operations level should be capable of delivering an emergency decon capability. At most "working" hazmat incidents, decon services will be provided by HMRTs or fire and rescue units working under the direction of a hazmat technician. Questions pertaining to disposal methods and procedures should be directed to environmental officials and technical specialists, based on applicable federal, state, and local regulations. See **FIGURE 4-14**.

**GOAL**: Ensure the safety of both emergency responders and the public by reducing the level of contamination on-scene and minimizing the potential for secondary contamination beyond the incident scene.

**FIGURE 4-14** Cleanup and recovery operations can extend well beyond the emergency response phase.

Courtesy of Gregory G. Noll.

CHECKLIST

- Ensure that the decon operations are coordinated with tactical operations. This should include the following tasks:
  - The decontamination area is properly located within the warm zone, preferably upslope and upwind of the incident location.
  - The decontamination area is well marked and identified.
  - The proper decontamination method and the type of personal protective clothing to be used by the decon team have been determined and communicated, as appropriate.
  - All decon operations are integrated within the ICS organization.
- Ensure proper decon of all personnel before they leave the scene. For example, flammable gases and some toxic and corrosive gases can saturate protective clothing and be carried into "safe" areas.
- Establish a plan to clean up or dispose of contaminated supplies and equipment before cleaning up the site of a release. Federal and state laws require proper disposal of hazardous waste.

*RESPONDER TIPS*

- Establishing an emergency decon capability should be part of the IAP for any incident where hazardous materials are involved.
- Permeation can occur with any porous material, not just PPE.
- Never transport contaminated victims from the scene to any medical facility without conducting field decon.

## Step 8: Terminate the Incident

**FUNCTION:** This is the termination of emergency response activities and the initiation of post-emergency response operations (PERO), including investigation, restoration, and recovery activities. This would include the transfer of command to the agency/facility that will be responsible for coordinating all post-emergency activities.

**GOAL:** Ensure that overall command is transferred to the proper agency/organization when the emergency is terminated and that all postincident administrative activities are completed per local policies and procedures.

CHECKLIST

- Account for all personnel before securing emergency operations.
- Conduct incident debriefing session for on-scene response personnel. Provide background information necessary to ensure the documentation of any exposures.
- Command is formally transferred from the lead response agency to the lead agency for all post-emergency response operations.
- Ensure that the following elements are documented:
  - All operational, regulatory, and medical phases of the emergency, as appropriate
  - All equipment or supplies used during the incident
  - The names and telephone numbers of all key individuals, including contractors, public officials, and members of the media
  - Required reports and documentation submitted in accordance with agency/facility policies and procedures
- Ensure that all emergency equipment is reserviced, inspected, and returned to proper locations. Provide a point of contact for all postincident questions and issues.
- Conduct a critique of all major and significant incidents based on local protocols.

*RESPONDER TIPS*

- Although every organization has a tendency to develop its own critique style, never use a critique to assign blame.
- Organizations must balance the potential negatives against the benefits that are gained through the critique process. Remember—the reason for doing the critique in the first place is to improve your operations.
- Most critiques fall into one of three categories: (1) We lie to each other about what a great job we just did, (2) we beat up each other for screwing up, or (3) we focus on the lessons that were learned and the changes/improvements that must be made to our response system.

## Summary

Emergencies involving flammable liquid bulk storage tanks and facilities have the potential to be long and complex response scenarios. Critical factors influencing incident complexity and duration will include the nature and location of the incident, available water and Class B foam resources, and the operational capabilities of emergency responders to handle large-scale flammable liquid emergencies.

The formula for a safe and effective response is having a coordinated incident command structure based upon the incident command system, and relationships that are developed and maintained well before the incident. Storage tank emergencies will bring a range of public and private sector players to the incident, including facility personnel;

emergency responders; environmental and emergency response contractors; and government agencies and officials from local, state, and federal levels.

Every incident must have an incident action plan (IAP) based upon the IC's assessment of (1) incident potential (i.e., visualizing hazardous materials behavior and estimating the outcome of that behavior), and (2) the initial operational strategy. Key elements in the development of the IAP will include the development of strategic goals (i.e., What are you going to do?), the establishment of tactical objectives (i.e., How are you going to do it?), and the implementation of task-level activities to achieve the tactical objectives (i.e., Do it). Response operations can encompass offensive, defensive, and nonintervention modes.

On-scene response operations should be based on a structured and standardized system of protocols and procedures. Regardless of the nature of the incident and response, a reliance on standardized procedures will bring consistency to the tactical operation and will help to minimize the risk of exposure and harm to all responders. This section uses the Eight Step Process© as the framework for reviewing the tactical-level functions and tasks.

## References and Suggested Readings

1. National Fire Protection Association, *NFPA Technical Standard 30—Flammable and Combustible Liquids Code.* Quincy, MA: National Fire Protection Association (2016).
2. Noll, Gregory G., and Michael S. Hildebrand, *Hazardous Materials: Managing the Incident* (4th edition). Burlington, MA: Jones and Bartlett Learning (2014).
3. Smith, James P., "History as a Teacher: How a Refinery Blaze Killed 8 Firefighters." *Firehouse Magazine* (December 1987), pp. 16–20.

# Firefighting Foam, Water Supply, and Fire Protection Requirements

CHAPTER 5

Courtesy of William T. Hand.

## Chapter Outline

- Objectives
- Key Terms
- Introduction
- Foam History
- Class B Firefighting Foams
- Storage Tank Safety and Fire Protection Systems
- Portable and Mobile Fire Protection Options
- Determining Foam Concentrate Requirements
- Firewater Supply and Delivery Systems
- Determining Water Supply Requirements
- Summary
- References

## Objectives

1. Describe the factors to be considered in evaluating and selecting Class B firefighting foam concentrates for use on flammable liquids (16.3.1).
2. Describe the factors to be considered for the portable application of Class B firefighting foam concentrates and related extinguishing agents for the following types of incidents (16.3.2):
   a. Flammable liquid spill (no fire)
   b. Flammable liquid spill (with fire)
   c. Flammable liquid storage tank fire
3. Given examples of different types of flammable liquid bulk storage tanks, identify and describe the application, use, and limitations of the types of fixed and semifixed fire protection systems that can be used, including the following (16.3.3):
   a. Foam chambers
   b. Catenary systems
   c. Subsurface injection systems
   d. Fixed foam monitors
   e. Foam and water sprinkler systems
4. Given the tank size and dimensions, determine the following (16.3.8):
   a. Applicable extinguishing agent(s)
   b. Approved application method (both portable and fixed system applications)
   c. Approved application rate and duration based upon NFPA 11 or other guidance used by the authority having jurisdiction.
   d. Required amount of Class B foam concentrate and required amount of water
   e. Volume and rate of application of water for cooling exposed tanks
5. Given the size, dimensions, and product involved for a flammable liquid bulk storage tank fire, determine the following (16.3.9):
   a. Applicable extinguishing agent(s)
   b. Approved application method (both portable and fixed system applications)
   c. Approved application rate and duration based upon NFPA 11 or other guidance used by the authority having jurisdiction
   d. Required amount of Class B foam concentrate and required amount of water
   e. Volume and rate of application of water for cooling involved and exposed tanks
   f. Recommendations for controlling product and water drainage and runoff
6. Given the size, dimensions, and product involved for a fire involving a single flammable liquid bulk storage tank and its dike area, determine the following (16.3.10):
   a. Applicable extinguishing agent(s)
   b. Approved application method (both portable and fixed system applications)
   c. Approved application rate and duration based upon NFPA 11 or other guidance used by the authority having jurisdiction

   d. Required amount of Class B foam concentrate and required amount of water
   e. Volume and rate of application of water for cooling involved and exposed tanks
   f. Recommendations for controlling product and water drainage and runoff
7. Given the size, dimensions, and product involved for multiple flammable liquid bulk storage tanks burning within a common dike area, determine the following (16.3.11):
   a. Applicable extinguishing agent(s)
   b. Approved application method (both portable and fixed system applications)
   c. Approved application rate and duration based upon NFPA 11 or other guidance used by the authority having jurisdiction
   d. Amount of Class B foam concentrate and water required
   e. Volume and rate of application of water for cooling involved and exposed tanks
   f. Recommendations for controlling product and water drainage and runoff

## Key Terms

*Class B Firefighting Foam* A firefighting foam designed to extinguish Class B fuels. Class B fuels can be subdivided into two subclasses: (1) hydrocarbons such as gasoline, kerosene, and fuel oil that will not mix with water, and (2) polar solvents such as alcohols, ketones, and ethers which will mix with water.

*Finished Foam* Final foam product as it exits at a discharge device (e.g., nozzle, monitor, foam chamber), after having been aerated.

*Firefighting Foam* A stable mass of small air-filled bubbles, which have a lower density than oil, gasoline, or water. Foam is made up of three ingredients: water, foam concentrate, and air. When mixed in the correct proportions, these three ingredients form a homogeneous foam blanket. Class B firefighting foam can be found in three specific physical phases: (1) foam concentrate, (2) foam solution (foam concentrate + water), and (3) finished foam product (i.e., foam solution + air).

*Foam Concentrate* Liquid concentrate that is mixed with water in the correct proportion to form a foam solution.

*Foam Solution* A solution of water and foam concentrate after they have been mixed together in the correct proportions. When properly aerated, the foam solution forms the firefighting foam as described above.

## Introduction

This chapter will discuss firefighting foam and water supply requirements for protecting, attacking, and extinguishing aboveground storage tank fires. We will also provide background information on fire protection options and requirements, including the selection, application, and use of Class B firefighting foam concentrates, the methods of determining foam concentrate requirements, fire protection systems found on storage tanks and in distribution terminals, types of water supply and delivery systems, and determining water supply requirements.

## Foam History

Class B firefighting foams are used to suppress flammable and combustible liquid spills and fires. Essentially, foam minimizes the production of vapors. Foam cools the fire and prevents fuel contact with air (oxygen), thereby resulting in the fire being extinguished. See FIGURE 5-1.

Foam as an extinguishing agent has been around for over 115 years. History credits the Russian engineer Aleksandr Loran with the invention of the first firefighting foam. Loran was a teacher in Baku, Azerbaijan, which remains a major oil production area. He observed how difficult large crude oil fires were to extinguish and set out to find a solution, which he did in 1902. The original foam was a mixture of two powders and water produced in a foam generator. Back in the day, it was called "chemical foam" because of the chemical reaction required to create the foam bubbles. The powders included sodium bicarbonate combined with aluminum sulfate, with other ingredients to keep the foam bubbles stable. The bubbles were generated by the production of carbon dioxide gas. The next time you visit a fire museum, you will likely see an antique chemical fire extinguisher. It was common in the fire apparatus days of the early 1900s in the United States for a fire department to have a "chemical unit." These were the

FIGURE 5-1 Class B Firefighting Operations.

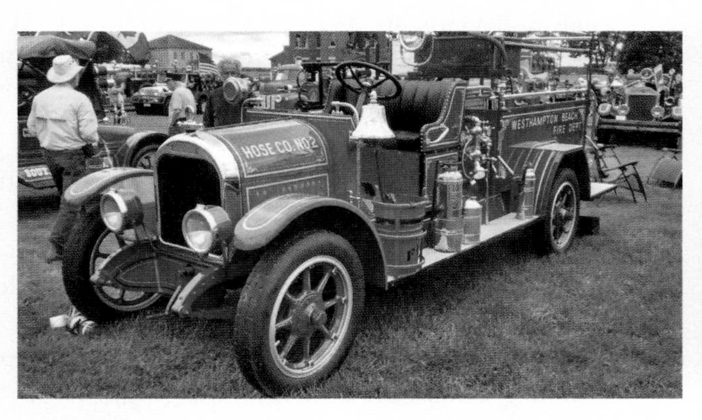

FIGURE 5-2 Class B firefighting units of the early 1900s were the frontline hazmat units.

New Jersey Fire Museum.

first hazmat response teams. Someone is always the first person to solve a problem! See FIGURE 5-2.

In the 1940s Percy Lavon Julian developed liquid protein foam made from soy protein. It was mixed with water in a proportioner or an aerating nozzle to make air bubbles. As the years went by another type of protein was substituted—ground and dissolved fish by-products (fish guts). If you ever smelled protein foam when it was applied you will never forget the odor; nevertheless, the stuff worked. If one could overlook the odors, it also made a great lawn fertilizer!

In the 1960s, National Foam developed fluoroprotein foam. Also in the 1960s, Dr. Richard Tuve of the U.S. Naval Research Laboratory developed aqueous film forming foam (AFFF), which was the first foam to have a dramatic knockdown capability on flammable liquid spills. This advancement was an important factor for aircraft firefighting, as AFFF demonstrated the ability to cover a spill and suppress vapor evolution. AFFF has also been combined with other extinguishing agents, such as Purple K (potassium bicarbonate) and ultrahigh-pressure (UHP) water for quick attack/rapid intervention scenarios such as those found in the aircraft firefighting community. See FIGURE 5-3.

In the 1970s polar solvents began to be used in a variety of manufacturing processes as well as a fuel source and gasoline additive. The AFFFs that were so effective on hydrocarbon fires were not effective on polar solvent (alcohol) fires. This led to the invention of alcohol-resistant AFFF (AR-AFFF) concentrates and eventually to the development of today's "3 × 3" foam concentrates such as National's Universal Foam and William's T-Storm products.

## Class B Firefighting Foams

In order for a flammable and combustible liquid to burn, four elements must be present: fuel, heat (ignition sources), air (oxygen), and an uninhibited chemical chain reaction. Under normal circumstances if any one of the elements is

FIGURE 5-3 The U.S. Air Force P-34 Rapid Intervention Vehicle combines the use of UHP water and AFFF.

U.S. Air Force.

removed or interfered with, the fire is extinguished. Firefighting foam does not interfere in the chemical reaction process.

Class B foam acts in the following manner:

- Suppresses the release of flammable vapors that can mix with air, which is critical in both fire and spill scenarios
- Smothers the fire by eliminating air
- Separates the flames/ignition source from the fuel surface
- Cools the fuel and any adjacent metal surfaces

From a chemical perspective, here are two basic categories of flammable liquid fuels:

**Standard Hydrocarbon Fuels**—These are products primarily made up of hydrogen and carbon that do not mix with water or are not miscible in water (i.e., these products all float on top of water). Examples include crude oils and refined petroleum products such as gasoline, diesel, kerosene, and jet fuel.

**Polar Solvent Fuels**—These fuels mix readily with water or are miscible in water. Examples include alcohols, ketones, and ethers.

Class B firefighting foam remains the workhorse for petroleum storage tank firefighting. It is used to extinguish fires involving flammable and combustible liquids and to suppress vapors from unignited spills. While other agents, such as dry chemicals (e.g., Purple K), are used for quick attack scenarios or to deliver the knockout punch for three-dimensional fires (e.g., a flange fire on a pipe), Class B foam remains the "weapon of choice" for large-scale flammable and combustible liquid problems.

Chemically, Class B foams can be divided into two general categories: synthetic based or protein based. Synthetic foams are basically super soap with fire performance additives. They include high expansion foam, AFFF, and AR-AFFF.

In general, synthetic foams are more fluid, flow more freely, and provide quick knockdown with limited postfire security or burnback resistance. Protein foams use natural protein foamers instead of a synthetic soap, and similar fire performance components are added. Protein-type foams include regular protein foam (P), fluoroprotein foam (FP), alcohol-resistant fluoroprotein foam (AR-FP), film forming fluoroprotein (FFFP), and alcohol-resistant film forming fluoroprotein (AR-FFFP). In general, protein-based foams are not as fluid and spread slightly slower than synthetic foams, but they can produce a more heat-resistant, longer lasting foam blanket.

## Selecting a Class B Foam

Selection of a foam concentrate is an important part of a successful firefighting operation. There are several different types of Class B foam concentrates sold by a variety of reliable manufacturers at different concentrations and for different fire protection applications. The selection of a foam concentrate can be as much a business decision as a technical one.

The selection of Class B foams for the protection of hazards found at aboveground petroleum storage tank facilities will be based upon the following criteria:

- Type of flammable liquid involved (i.e., hydrocarbon or polar solvent) (As noted, some foam concentrates are not suitable for use on polar solvent/alcohol type spills and fires.)
- Type and nature of the hazard being protected (i.e., storage tank, loading rack, pump stations, spill vs. fuels in-depth)
- Overall size and area of the hazard to be protected

The most common foam concentrates used for flammable liquid storage tank fire protection are as follows:

**Fluoroprotein (FP) Foam**—Combination of protein-based foam derived from protein foam concentrates and fluorochemical surfactants. The addition of the fluorochemical surfactants produces a foam that flows easier than regular protein foam. Fluoroprotein foam can also be formulated to be alcohol resistant.

Key characteristics of FP foam as it relates to tank firefighting include the following:

- Available in 3% and 6% concentrations
- Is oleophobic (it sheds oil) and can be used for subsurface injection
- Compatible with the simultaneous application of dry chemical extinguishing agents (e.g., Purple K)
- Forms a strong foam blanket for long-term vapor suppression on unignited spills
- Must be delivered through air aspirating equipment
- Has a shelf life of approximately 10 years

**Aqueous Film Forming Foam (AFFF)**—AFFF is a synthetic foam consisting of fluorochemical and hydrocarbon surfactants combined with high boiling point solvents and water. AFFF film formation is dependent upon the difference in surface tension between the fuel and the firefighting foam. The fluorochemical surfactants reduce the surface tension of water to a degree less than the surface tension of the hydrocarbon so that a thin aqueous film can spread across the fuel. AFFF was the first foam developed that could be flowed through nonaspirating structural firefighting nozzles.

Key characteristics of AFFF as it relates to tank firefighting include the following:

- Available in 1%, 3%, and 6% concentrations for use with either fresh or salt water
- Very effective on spill fires with a good knockdown capability
- Compatible with simultaneous application of dry chemical extinguishing agents (e.g., Purple K)
- Suitable for subsurface injection

**Alcohol-Resistant AFFF (ARC)**—Alcohol-resistant AFFFs are commonly available at 3% hydrocarbon: 6% polar solvent (known as 3 × 6 concentrate) or 3% hydrocarbon: 3% polar solvent (known as 3 × 3 concentrate) options. When applied to a polar solvent fuel, they will often create a polymeric membrane rather than a film over the fuel. This membrane separates the water in the foam blanket from the attack of the polar solvent. Then, the blanket acts in much the same manner as a regular AFFF. ARC is the overall recommended agent when dealing with reformulated gasoline and ethanol blends. (See Scan Sheet 5-B later in the chapter for more details on reformulated gasoline fire suppression recommendations.)

Key characteristics of alcohol-resistant AFFF as it relates to tank firefighting include the following:

- Must be applied gently to polar solvents so that the polymeric membrane can form first
- Should not be plunged into the fuel but gently sprayed over the top of the fuel
- Very effective on spill fires with a good knockdown capability
- May be used for subsurface injection applications

**Film Forming Fluoroprotein (FFFP) Foam**—Based on fluoroprotein foam technology with AFFF capabilities, FFFP foam combines the quick knockdown capabilities of AFFF along with the heat resistance benefits of fluoroprotein foam.

Key characteristics of FFFP foam as it relates to tank firefighting include the following:

- Available in 3% and 6% concentrations
- Compatible with simultaneous application of dry chemical extinguishing agents (e.g., Purple K)
- Can be used with either fresh or salt water

FFFP foam is also available in an alcohol-resistant formulation. This type has all of the properties of regular FFFP foam, as well as the following characteristics:

- Can be used on hydrocarbons at 3% and polar solvents at 6%; newer FFFP concentrates can be used on either type of fuel at 3% concentrations
- Can be used for subsurface injection
- Can be plunged into the fuel during application

# Scan Sheet 5-A—Properties, Comparisons of Firefighting Foam Types, and Compatibility

| Property | Protein | Fluoroprotein | AFFF | FFFP | AR-AFFF |
|---|---|---|---|---|---|
| Knockdown | Fair | Good | Excellent | Good | Excellent |
| Heat Resistance | Excellent | Excellent | Fair | Good | Good |
| Fuel Resistance* | Fair | Excellent | Moderate | Good | Good |
| Vapor Suppression | Excellent | Excellent | Good | Good | Good |
| Alcohol Resistance | None | None | None | None | Excellent |

*Hydrocarbons

A few rules should be observed regarding Class B foam compatibility:

1. Similar foam concentrates by different manufacturers are not considered to be compatible in storage applications (e.g., different foam concentrates should not be mixed together in the same storage tank). The exception to this would be Mil-Spec (i.e., military specification) foam concentrates. The Mil Spec standards are written so that mixing of foam concentrates can be done with no adverse effects.
2. Don't mix different kinds of foam concentrates (e.g., AFFF and fluoroprotein) before or during proportioning.
3. On the emergency scene, concentrates of a similar type (e.g., all AFFFs, all fluoroprotein) but from different manufacturers may be mixed together immediately before application.
4. Finished foams of a similar type but from different manufacturers (e.g., all AFFFs) are considered compatible.

# Scan Sheet 5-B—GASOLINE and ETHANOL BLENDS

Under the requirements of the U.S. EPA Renewable Fuels Standard Program, nearly all gasoline sold as a motor fuel in the United States contains up to 10% fuel grade ethanol in regular unleaded gasoline. This 10% blend is known as E-10, indicating that a total volume of the gasoline/ethanol blend mixture is 10% alcohol.

© Jones & Bartlett Learning.

© Jones & Bartlett Learning.

With regulatory action calling for the increased blending of ethanol with hydrocarbon fuels, there will be an increase in the type of mixtures found. Current formulations and mixtures include the following:

- Ethyl alcohol (100% ethanol by volume) or E100, which is shipped under UN1170
- Denatured alcohol (95%–99% ethanol by volume); includes E95–E98, which are shipped under UN1987
- E10 and lower concentrations (1%–10% ethanol by volume); include E10 unleaded gasoline, shipped under UN 1203
- Ethanol mixtures of 11%–94% ethanol by volume; include E85 and Ethanol Flex Fuel, shipped under UN3475

The Renewable Fuel Standards have mandated increases in the use of ethanol. The automobile industry in turn has introduced more flexible fuel vehicles that will demand more of these fuels at the pump.

The most common ethanol blend being transported today by rail tank car is E95, or denatured ethanol (UN 1987). It may be delivered directly to a petroleum terminal if the facility has the ability to accept rail traffic. If not, it will go to a railyard or siding that is designed to transload the product from the railcar to a highway cargo tank truck. From this facility, it will then travel to a petroleum terminal where it will be offloaded into separate bulk storage tanks until it is mixed with gasoline, at the appropriate percentage, at the cargo tank truck loading rack. Denatured ethanol may also be transported by barge or ship. The shipment of ethanol via pipeline is extremely limited because of the corrosivity of the ethanol. Note: For more detailed information on gasoline tank trucks, see Hildebrand, Noll, and Hand's *Gasoline Tank Truck Emergencies* (4th edition, Jones and Bartlett Learning, 2016).

Emergency responders are generally familiar with the characteristics of gasoline, its hazards, and how to extinguish flammable liquid fires. There are some similarities between gasoline and ethanol, as both are organic solvents and are highly volatile. The primary differences between the two fuels are that ethanol is a polar solvent (i.e., completely soluble in water) and gasoline is a hydrocarbon (i.e., low solubility in water). As a result, the selection and use of Class B firefighting foams is critical, as regular Class B firefighting foams are *not* effective on polar solvents. Both products are toxic by inhalation and present the risk of fire in open unconfined areas. Ethanol has a slightly wider flammable range than gasoline. The following chart provides a comparison of the two fuels.

### Hazards of Ethanol vs. Gasoline

| PRODUCT | FLASH POINT | BOILING POINT | LFL | UFL | IDLH |
|---|---|---|---|---|---|
| Gasoline | −45°F (−42.8°C) | 100°F (37.8°C) to 400°F (204.4°C) | LFL = 1.4% | UFL = 7.4% | 500 ppm |
| Ethanol | 55°F (12.8°C) | 173°F (78.3°C) | LFL = 3.3% | UFL = 19.0% | 3,300 ppm |

Based on a series of performance fire tests conducted by the American Petroleum Institute (API) and the Ethanol Emergency Response Coalition (EERC), tactical recommendations for dealing with gasoline–ethanol mixtures can be summarized as follows:

- For fires involving gasoline–ethanol mixtures up to 10% will behave like a hydrocarbon fire, whereas 10% to 15% will assume more of the burning properties of a polar solvent.
- Regular AFFF and AR-AFFF/ARC will be effective for these mixtures, although an increased application rate will be required when using regular AFFF, especially for those scenarios requiring prolonged burnback resistance.
- Regular Class B firefighting foams (e.g., AFFF, fluoroprotein foam) will not be effective for either ethanol or gasoline–ethanol mixtures greater than 10% ethanol.
- Overall, alcohol-resistant AFFF (AR-AFFF or ARC) is the best agent for dealing with both ethanol and gasoline–ethanol blends.

*Sources:* Ethanol Emergency Response Coalition, *Quick Reference Guide: Fuel Grade Ethanol Spills* (Including E-85); Evaluating the Use of Firefighting Foam, *Fire Engineering* (February 1986), pages 44–49.

## Storage Tank Safety and Fire Protection Systems

Bulk storage facilities often include engineered fire protection systems that can provide emergency responders with numerous options for spill and fire control. Understanding how these systems function and their limitations could mean the difference between a good and bad outcome. Emergency responders should be familiar with the basic principles of operation of fixed and semifixed fire protection systems and understand how they may be used to support offensive and defensive firefighting operations.

Based on the risks present and the requirements of the authority having jurisdiction (AHJ), fire protection systems at bulk distribution facilities can include the following:

- Diesel-, electric-, gas-, or steam-powered fire water pumps
- Foam concentrate storage tanks to supply fire protection systems
- Detection and fixed fire protection systems at loading racks
- Fixed or semifixed foam suppression systems on storage tanks
- Spill containment at loading racks and around transfer pumps
- Water spray or sprinkler systems for exposure protection of critical areas
- Fireproofing to protect critical vertical and horizontal support structures

### Fixed and Semifixed Fire Protection Systems

Flammable liquid storage tanks and facilities can be protected with a variety of installed fire protection systems. *NFPA 11—Low, Medium and High Expansion Foam Extinguishing Systems* categorizes storage tank fire protection systems as fixed and semifixed:

- **Fixed systems** are complete installations piped from a central foam station and discharge finished foam through fixed delivery outlets onto the hazard to be protected. If a pump is required, it is usually permanently installed. Fixed systems are activated either manually or automatically by a fire detection system. Fixed systems are commonly used to protect flammable liquid loading racks. When found, fixed systems on storage tanks will likely have longlasting or unlimited water supply and large quantities of foam concentrate.

- **Semifixed systems** are those where the hazard is equipped with a fixed discharge outlet connected to piping, which terminates a safe distance from the hazard. Emergency responders must then supply the necessary foam solution by connecting to either piping or fire department connections away from the hazard area.

There are several types of installed fire protection systems found at storage tank facilities. These are discussed next.

## Foam Chambers

Foam chambers are fixed discharge outlets attached to the outside tank shell that apply foam directly onto the surface of the burning fuel. They are commonly used for the protection of cone roof, open floating roof, and covered floating roof tanks. (See FIGURE 5-4A & B.)

Foam chambers are installed at equally spaced positions on the tank shell just below the roof joint. The number of foam chambers required for a tank is based upon the tank diameter and is described in NFPA 11. For example, a tank 80 to 120 ft. (24 to 36 m) in diameter would require two discharge outlets; a tank 180 to 200 ft. (54 to 60 m) in diameter would require six discharge outlets.

Foam chambers are connected to piping, which transports the foam solution from the proportioning source outside the dike wall to the foam chamber. They may be supplied from either a fixed or semifixed system arrangement, although semifixed systems are most common. Frangible seals at the discharge outlet of the foam chamber prevent vapors from entering the foam piping. These seals are designed to burst when foam pressure is applied. The foam chamber also contains air inlets to aerate the foam solution, an expansion area, and a discharge deflector. The deflector is located inside the tank shell, so that foam flows against the inside shell and gently onto the fuel surface.

On open floating roof tanks, foam chambers are designed to protect only the seal area. If the floating roof sinks or tilts, a full surface fire will result. If this occurs, "over-the-top" applications from portable application devices and/or mobile fire apparatus will be required. On covered floating roof tanks with steel double deck roofs or pontoon floating roofs, the foam chamber system can be designed for only the protection of the seal area. All other types of internal floating roof tanks (e.g., pan) must be protected against a full surface fire.

Foam chamber systems designed to protect the seal area on floating roof tanks must also have a foam dam installed on the roof. The foam dam retains the foam over the seal or weather shield. It is normally 12 or 24 inches (300 or 600 mm) high, and should be located at least 1 ft. (0.3 m) but no more than 2 ft. (0.6 m) from the tank wall. Where a secondary seal is installed, the foam dam should extend at least 2 inches (50 mm) above the top of the secondary seal.

FIGURE 5-4A & B Foam chambers can deliver foam directly onto the surface of the burning fuel.

Refinery Terminal Fire Company.

Fires involving large diameter or "jumbo" storage tanks (i.e., greater than 250 ft. diameter) pose unique challenges, even when protected with fire protection systems. One way of meeting the challenge is through the use of a fixed or portable tank-mounted foam nozzle installed just below the top rim of the tank in place of the foam chamber. The foam lands on the fuel surface and flows to the center of the tank, which overcomes the problem of getting foam to flow across the wide surface of a large diameter tank. (See FIGURE 5-5.)

Advantages of foam chambers include the following: each system is engineered specifically for a particular application, and not as much foam is lost or wasted when compared to "over the top" portable application devices. The primary disadvantage is that these systems may be damaged by the initial fire or explosion.

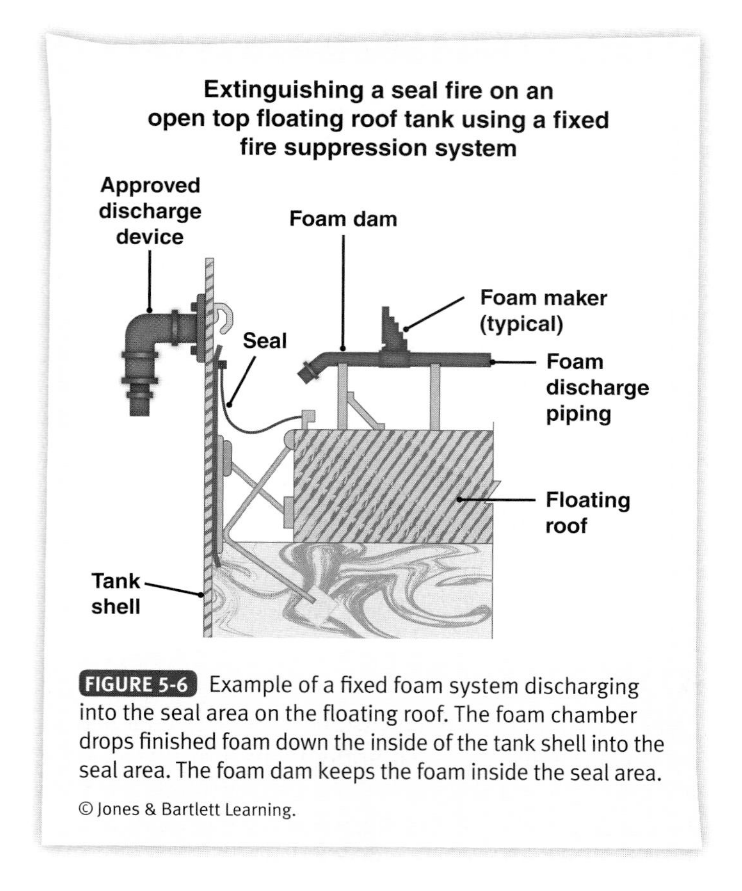


FIGURE 5-6 Example of a fixed foam system discharging into the seal area on the floating roof. The foam chamber drops finished foam down the inside of the tank shell into the seal area. The foam dam keeps the foam inside the seal area.



## Foam Discharge on Open-Top Floating Roof

Open-top floating roof tanks can be protected with foam discharge systems located on the roof itself. These systems can either provide foam onto the top of the seal or supply foam beneath the seal, and may be either fixed or semifixed systems. (See FIGURE 5-6.)

- **Top-of-Seal Protection (aka Catenary System).** This is basically the same as described for foam chambers located on the tank shell. The only significant difference is that the foam discharges and the associated piping are located on the floating roof. This piping is supplied by a feed line that either runs up the side of the tank and down the stairway or through the inside of the tank to the underside of the floating roof. A flexible hose is used near the top of the system to allow for the movement of the roof. A circle of piping then follows around the edge of the floating roof and is connected to the foam makers. Tanks protected by top-of-seal protection must have foam dams.
- **Below-the-Seal Protection (aka Coflexip Method).** This is the same design as top-of-seal systems. The primary difference is that the foam discharge orifice actually penetrates the seal, thereby allowing foam to be applied directly to the fuel surface.

FIGURE 5-5 The Daspit Wand provides the ability to apply foam onto the center of the tank surface through the use of a portable foam monitor applied onto the tank shell.

Refinery Terminal Fire Company.

## Subsurface Injection Systems

Subsurface injection systems inject foam into the base of the storage tank and allow the foam to "float" to the top of the fuel where it forms a foam blanket over the surface of the fuel. Foam that is injected subsurface is expanded less than foam that is applied through surface or portable applications (e.g., 4:1 versus 20:1 or higher).

The foam may be discharged into the tank through either separate foam delivery lines or through the tank's product fill line. If separate foam lines are used, they must be spaced equally around the edge of the tank. It is important that the foam be discharged above the layer of water commonly found resting in the bottom of the tank. This water collects as a result of both condensation and leaks; attempting to pump foam through this water layer will result in the destruction of the foam.

Most subsurface injection systems are semifixed systems. Because of the amount of piping involved in these systems and the backpressure of the fuel created by the column of fuel (weight of the fuel pushing downward) in the storage tank, a high-back pressure foam maker is required. A high-back pressure foam maker is an inline aspirator used to deliver foam under pressure. Air is supplied directly to the foam solution through a venturi action, which results in a low air content but in homogeneous and stable foam.

Fluoroprotein, AFFF, and FFFP foams are most commonly used for subsurface injection. However, regular protein foams cannot be used, as the foam will become saturated with the flammable liquid and burn after it rises to the surface. Subsurface injection systems cannot protect tanks containing polar solvents, as the foam would be destroyed by the fuel before reaching the fuel surface. Subsurface injection also cannot be used for hydrocarbon products with a viscosity above 2,000 SSI at 60°F (15°C), such as Bunker C oil and asphalt, or any fuel heated above 200°F (93°C). When used on floating roof tanks, the position of the floating roof (e.g., tilted, sunk) will be a critical variable in the success of the subsurface foaming operation.

Two primary advantages of subsurface injection systems are that the foam is efficiently delivered to the fuel surface without being affected by wind or thermal updrafts, and the lower probability of subsurface foam equipment being damaged by the initial fire or explosion. (See **FIGURE 5-7**.)

## Semisubsurface Injection Systems

The equipment used for these systems is basically the same as that used for regular subsurface injection systems. The primary difference is the actual delivery point of the foam. Semisubsurface injection systems discharge the foam through a flexible hose that rises from the bottom of the tank up through the fuel to the surface.

Under nonfire conditions, this flexible hose is contained within a housing located at the base of the tank. When the foam system is placed in service, the hose is released from the housing. The buoyancy of the foam inside the hose causes the hose to float to the surface of the fuel. Once deployed, the hose discharges foam directly to the surface of the burning fuel.

Semisubsurface injection systems have the same advantages as subsurface systems with one significant addition. Because semisubsurface injection systems actually apply foam directly onto the fuel surface, they can be used on tanks containing polar solvents provided alcohol-resistant concentrates are used. (See **FIGURE 5-8**.)

## Fixed Foam Monitors

Small flammable and combustible liquid tanks, as well as loading racks and dike areas, may be protected with foam

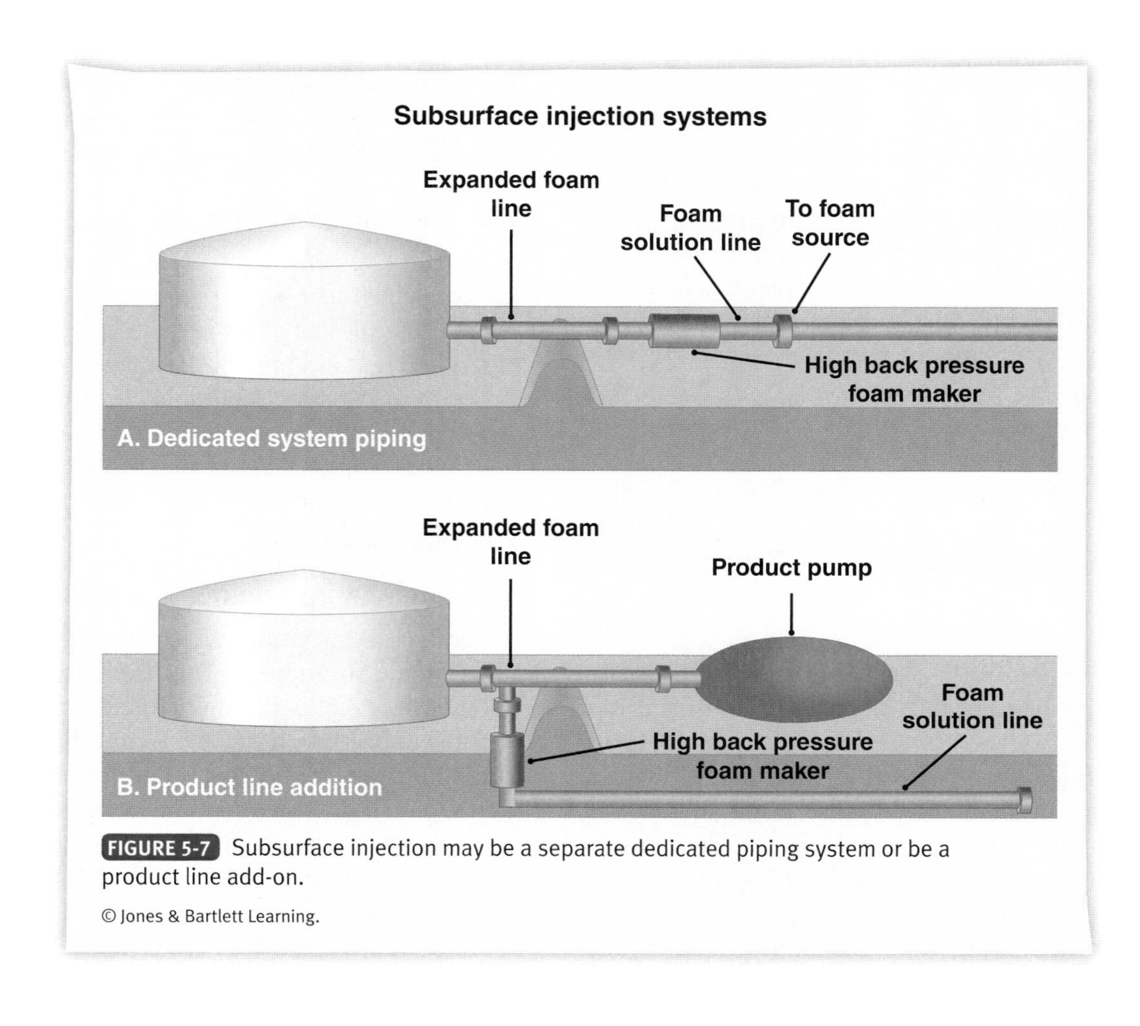


**FIGURE 5-7** Subsurface injection may be a separate dedicated piping system or be a product line add-on.

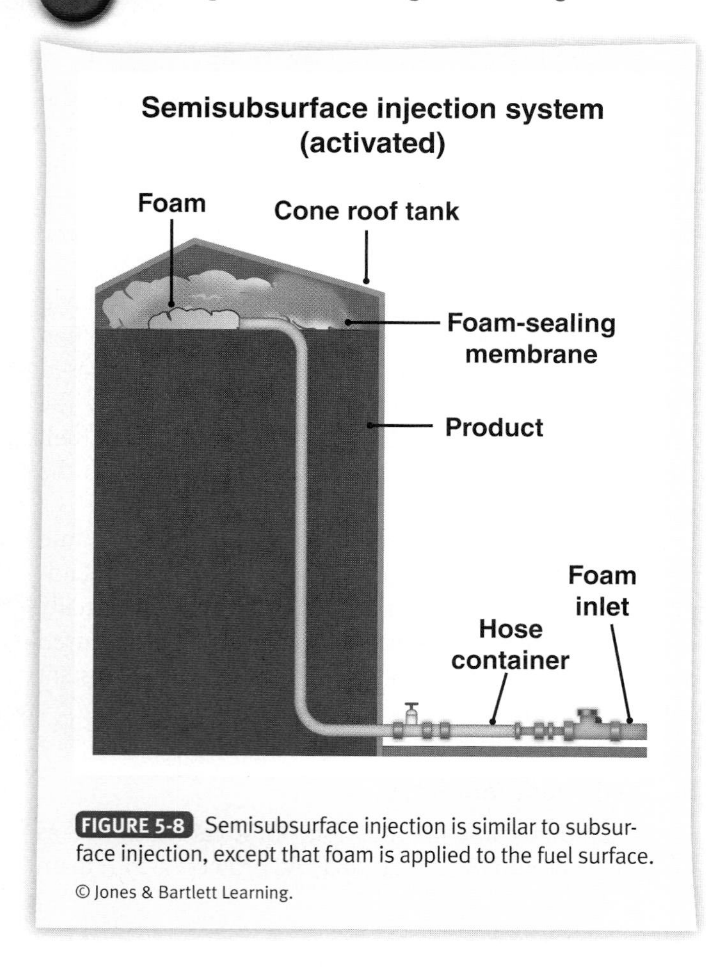


FIGURE 5-8 Semisubsurface injection is similar to subsurface injection, except that foam is applied to the fuel surface.

© Jones & Bartlett Learning.

FIGURE 5-9 Fixed fire monitors provide the advantage of being quickly placed in service.

Courtesy of Gregory G. Noll.

monitors and/or discharge connections for foam handlines. (See FIGURE 5-9.)

According to NFPA 11, foam monitors may be used for the protection of fixed roof tanks up to 60 ft. (18 m) diameter. Monitors operated at grade level are usually not recommended for floating roof seal fire protection because of the difficulty of directing foam onto the annular space and the danger of sinking the roof.

Advantages of fixed foam monitors include lower costs and less potential for damage in the initial stages of an incident. Disadvantages include the need to potentially expose personnel to greater hazards in close proximity of the tank to initiate operation of the monitor and susceptibility of foam streams to thermal updrafts and winds.

## Fixed Foam Handline Systems

Handlines may be used for the protection of fixed roof tanks not over 30 ft. (9 m) in diameter or 20 ft. (6 m) in height. Seals on open floating roof tanks less than 250 ft. (76 m) in diameter may also be protected with handline systems. These tanks require a foam dam like those previously described. The foam system consists of a riser that extends up the outside of the tank shell to the area near the top of the stairway. A single, fixed chamber is required at the top of the stairs to provide a safe base of operations for personnel operating in that area. The fixed outlet (i.e., standpipe connection) must be able to flow at least 50 gpm. Two 1.5-inch discharge outlets are typically provided at the top of the stairs to connect the handlines.

Advantages of fixed foam handlines include lower costs and ease of use. Disadvantages include the need to potentially expose personnel to greater hazards and susceptibility of foam streams to thermal updrafts and winds. (See FIGURE 5-10.)

## Fixed Low-Level Foam Discharge Outlets

Dike areas are sometimes protected by fixed foam makers connected to a fixed piping system around the dike area. These discharge devices are equally spaced around the dike area and are similar in design to those found on open-top floating roof tanks. Low-level foam systems may be either fixed or semifixed systems, and are typically found for protecting storage tank risks at industrial facilities as compared to petroleum distribution terminals. (See FIGURE 5-11.)

## Foam/Water Sprinkler Systems

These are special sprinkler systems designed and installed with a foam solution delivery capability. These systems provide an excellent foam blanket and consistent foam expansion because of preengineered characteristics. They provide coverage similar to that of standard sprinklers, although directional or specialty-type discharge devices may be used for specific applications. There are four types of foam sprinkler systems: wet pipe, dry pipe, preaction, and deluge systems. Foam sprinklers are commonly used for the protection of flammable liquid loading racks.

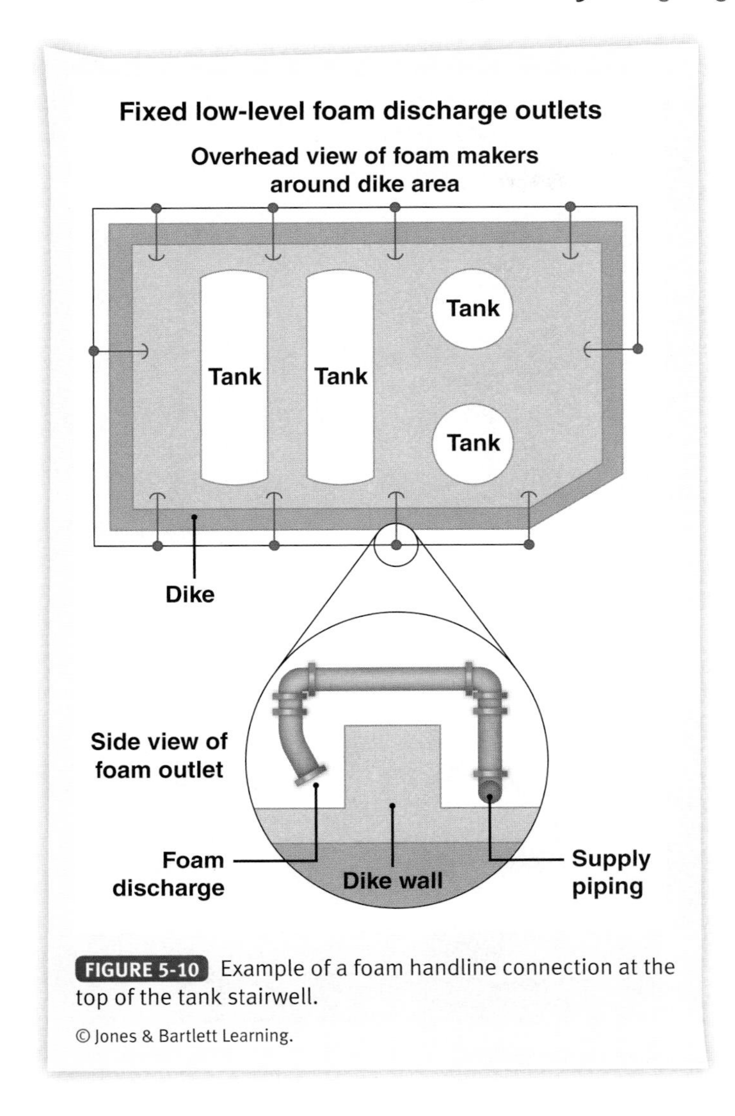


FIGURE 5-10 Example of a foam handline connection at the top of the tank stairwell.



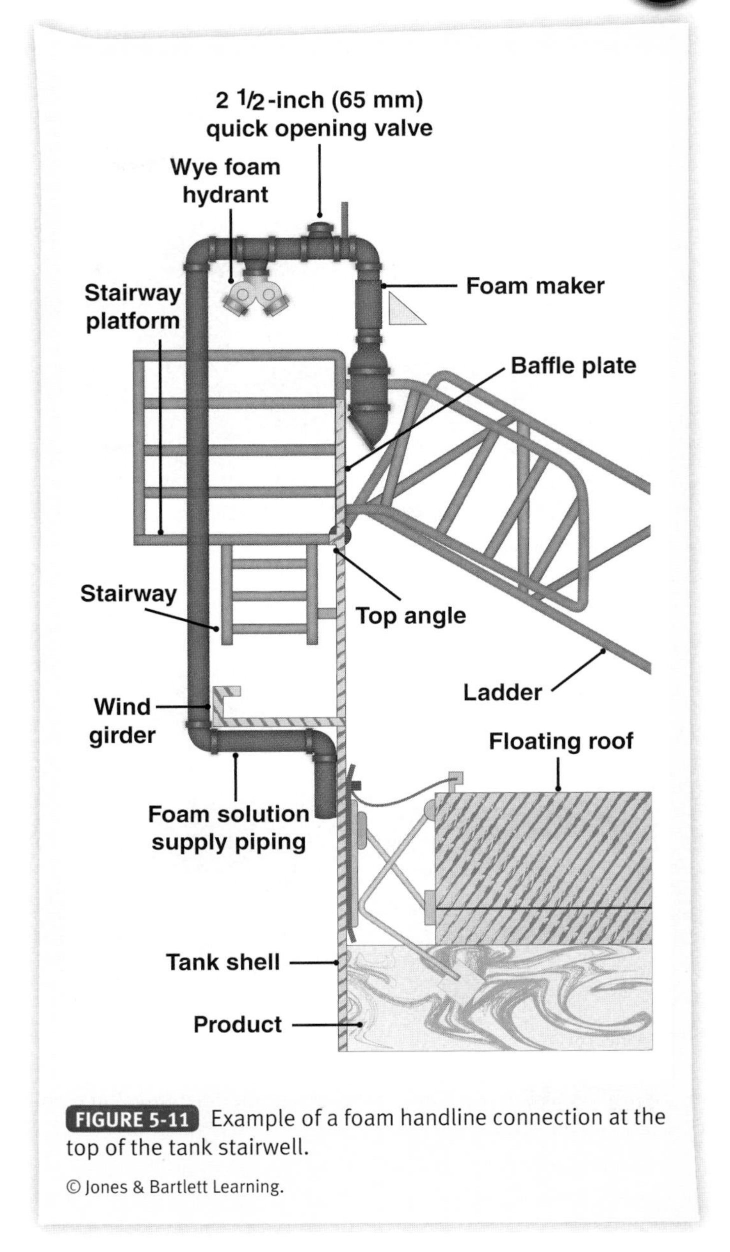


FIGURE 5-11 Example of a foam handline connection at the top of the tank stairwell.



## Portable and Mobile Fire Protection Options

When storage tanks are not protected by fixed and semi-fixed systems, emergency responders must then rely upon manual firefighting efforts to control and extinguish the fire. The following is an overview of the options available and their tactical application in the field.

### Portable Foam Monitors and Handlines

Portable foam monitors capable of flowing 5,000 gpm (18,927 g/L) or greater are commonplace in many petroleum industry fire departments. Large diameter storage tanks created the demand for portable foam monitors on trailers capable of flowing 10,000 to 12,000 gpm (37,800 to 45,424 L) of water or foam solution. Thirty years ago achieving this high-volume flow capacity from a single portable monitor was thought to have been impossible; today, it is standard firefighting equipment in many refineries. These improvements in large-volume foam and water movement technology have also led to increased performance and effectiveness in controlling and extinguishing storage tank fires through portable applications. (See FIGURE 5-12.)

Large-volume portable monitors (e.g., DASPIT Tool) see Figure 5-5 that can be attached to any strong point along I-beams, tank rims, crosswalks, etc., have also been an effective tool in scenarios such as a sunken floating roof, rim seal fire, and vapor suppression/mitigation. These monitors can be easily set up with one or two firefighters with flows ranging from 1,250 to over 2,000 gpm.

Foam handlines may be used for the protection of fixed roof tanks not over 30 ft. (9 m) in diameter or 20 ft. (6 m) in height. They are often used for the control and extinguishment of ground and dike fires, as well as for seal fires on open-top floating roof tanks.

Critical success factors in relying upon portable foam application devices include ensuring that sufficient foam concentrate, water, proportioning, and application devices are on hand and in place before the operation is initiated. As foam will be applied "over the top" of the tank, the incident commander must ensure

Courtesy of Gregory G. Noll.

Courtesy of Gregory G. Noll.

Courtesy of Gregory G. Noll.

Refinery Terminal Fire Company.

**FIGURE 5-12** A, B, C, and D Trailer-mounted, large-volume portable foam monitors are a common resource in the petroleum refining industry.

that responders have the capability to apply a foam that can overcome the thermal updraft as well as winds and distance. Remember, it is not how much foam or water that is thrown at the fire, but how much actually reaches the surface of the burning fuel!

Historically, emergency responders relied upon the "surround and drown" philosophy for storage tank fire extinguishment. Simply stated, portable application devices of various types and sizes were placed around the storage tank and a foam attack was initiated. The results could be characterized as "hit or miss"—sometimes it worked, but most often it did not. Over the last two decades, however, there have been significant improvements in both the equipment and body of knowledge for extinguishing storage tank fires. Among this body of knowledge are the following techniques:

- **The Window™**—The burning pattern/dynamics of a storage tank fire are reviewed by onscene personnel and the area where the fire is drafting air is identified (known as "The Window"). Once identified, all of the foam streams are then concentrated and directed as a massed application through "The Window" and onto the fire. The result of this action is the application of a larger concentration of foam onto the fuel surface. This technique has been successfully employed by Williams Fire & Hazard Control, Inc. for the extinguishment of a number of large diameter storage tank fires.
- **The Footprint™**—The "footprint" is the area where the foam lands on the surface of the burning fuel. Nozzle reach, pattern, and foam expansion are critical elements in developing a quality foam that will reach the fuel surface. Newer generation foam nozzles and foam concentrates (e.g., AR-AFFF) are used to create low expansion foams which are not as easily destroyed by the fire's thermal updrafts or affected by the wind generated by the fire. As a result, more foam reaches the surface of the fire.

## Portable Foam Wands

These are designed to deliver foam where logistics or safety considerations require that the foam nozzle be "unattended." Foam wands are often used in hard to access areas and can complement other foam application methods. Flow rates can range from 60 to 125 gpm (227 to 473 pm). The foam wand works in conjunction with an inline foam eductor or other foam proportioning system.

Foam wands are designed to hang on the tank shell and be supplied by foam lines. The size and weight of foam wands can vary greatly based upon the size and nature of the tank fire. For example, lightweight foam wands can be manually positioned and placed in service, while larger and heavier foam wands may require the use of aerial apparatus or cranes to be positioned.

## Mobile Foam Apparatus

Mobile foam apparatuses can be found at airports, petroleum and petrochemical industrial facilities, and in some municipal fire departments. These apparatuses use various combinations of foam proportioning systems, discharge devices, and foam concentrates, based upon the types of hazards they are designed to protect. Options include the following:

1. **Aircraft Rescue and Firefighting Apparatus (ARFF).** ARFF vehicles are designed to provide immediate suppression and extinguishment of flammable liquid spills and fires on airport properties. However, they often respond off airport property to provide mutual aid assistance to municipal fire departments at large-scale flammable liquid emergencies. Apparatus categories include major firefighting vehicles (i.e., crash trucks), rapid intervention vehicles, and combined agent vehicles.

   When evaluating the application and use of ARFF vehicles at storage tank emergencies, the IC must recognize both the application and the limitations of the unit. NFPA 11 recommends an application duration of 15 minutes for flammable liquid spill fires and 65 minutes for flammable liquid storage tank fires. While ARFF units may provide the ability to quickly knock down or overwhelm some surface spill fires, experience shows that they have limited success on storage tank fires.
2. **Industrial Foam Pumpers.** Many refineries and petrochemical facilities are equipped with large-capacity foam pumpers. Although comparable to their cousins with the municipal fire service, industrial foam pumpers are primarily intended to produce large water flows and quantities of Class B foam. (See FIGURE 5-13.)

   Most industrial foam pumpers are equipped with fire pumps that range in capacity from 1,500 to over 3,000 gpm, and foam tanks ranging from 500 to 1,500 gallons. Typical fixed mounted foam monitors range from 1,500 to 3,000 gpm. Specialized pumpers capable of flowing as much as 7,500 to 10,000 gpm can also be found.

   Many industrial apparatuses are also equipped with large fixed foam/water monitors capable of flowing the capacity of the fire pump. Supply lines as large as 7 inches are not uncommon. Some facilities also have aerial apparatus with similar capabilities. For example, an industrial fire department in South Texas has an aerial device capable of flowing over 5,000 gpm at an elevation of 110 ft!

   Because of their unique design and construction, as well as their ability to flow large quantities of foam and water, industrial fire apparatuses are well suited for the requirements of storage tank firefighting.
3. **Municipal Fire Apparatus Equipped with Foam Systems.** Some municipal fire departments also equip their pumpers with a Class B foam system. Around-the-pump proportioners, which provide foam to all discharges on the apparatus, are commonly found. Most foam tanks range from 20 to 100 gallons of foam concentrate.

Although there are some exceptions, most municipal fire apparatuses equipped with a foam system are designed to handle smaller scale flammable liquid emergencies, such as those found at vehicle accidents, industrial fires, and so forth.

# Determining Foam Concentrate Requirements

The availability of water and foam concentrate is a critical factor in evaluating the risks involved in a storage tank emergency. The IC must evaluate the following factors in developing the incident action plan:

1. Size of the fire (i.e., area involved, spill fire, tank fire, combination tank and dike fire)
2. Type of fuel (i.e., hydrocarbon or polar solvent)
3. Required foam application rate
4. Amount of foam concentrate required onscene and the ability to resupply it
5. Ability to deliver the required amount of foam/water onto the fuel surface and sustain the required flow rates

If an adequate water and foam supply is not available for both protecting exposures and controlling the fire, the IC should consider implementing defensive or nonintervention tactics until sufficient resources are available. As a general tactical guideline, foam application operations should not be initiated until sufficient foam concentrate is onsite to extinguish 100% of the exposed flammable liquid surface area.

Refinery Terminal Fire Company.

Courtesy of Monroe Energy.

Courtesy of Michael S. Hildebrand.

Courtesy of Michael S. Hildebrand.

**FIGURE 5-13** A, B, C, and D Examples of industrial foam fire apparatuses used in the petroleum industry.

## NFPA Recommended Foam Application Rates

In evaluating specific types of Class B foam concentrate for the protection and/or extinguishment of specific fire scenarios (e.g., spill, tank fire, hydrocarbon vs. polar solvent), emergency responders should always review the technical data package and the minimum foam application rates published by the respective foam manufacturer. (See **FIGURE 5-14**.)

The NFPA 11–recommended minimum foam application rates for specific fuels, foams, and applications are as follows:

- 0.10 gpm/ft$^2$—fixed system application for hydrocarbon fuels (e.g., cone roof storage tank with foam chambers).
- 0.30 gpm/ft$^2$—fixed system application for seal protection on an open-top floating roof tank.
- 0.10 gpm/ft$^2$—subsurface application for hydrocarbon fuels in cone roof tanks.
- 0.10 gpm/ft$^2$ (AFFF, FFFP) to 0.16 gpm/ft$^2$ (protein, fluoroprotein)—portable application for hydrocarbon spills (e.g., 1.75-inch handlines with foam nozzles).
- 0.16 gpm/ft$^2$—portable application for hydrocarbon storage tanks (e.g., portable foam cannons and master stream devices). A foam application rate of 0.18 to 0.20 gpm/ft$^2$ has been used to successfully extinguish hydrocarbon fires in large diameter tanks using master stream portable nozzles.
- 0.20 gpm/ft$^2$—*minimum* recommended rate for polar solvents. Higher flow rates may be required depending on the fuel involved and the foam concentrate used.

## Determining Foam Requirements

Foam concentrate requirements can be determined by the following process:

1. *Determine the type of fuel involved—hydrocarbon or polar solvent. This will determine the type of foam concentrate to be used.*
2. *Determine the surface area involved.*
   - Calculate Storage Tank Area = (0.785)(diameter$^2$) or simply round out to (0.8)(diameter$^2$)

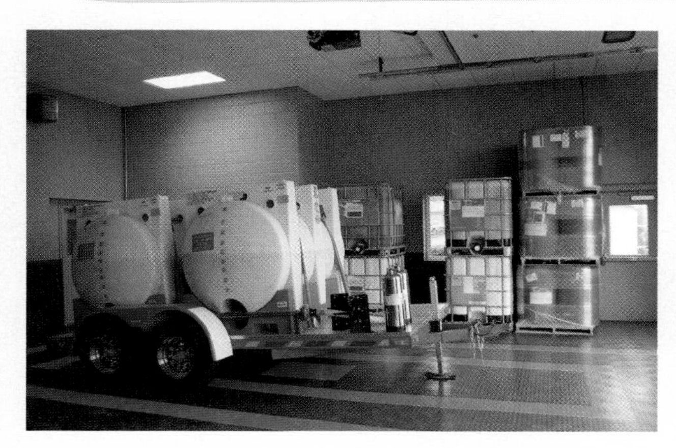

Courtesy of Gregory G. Noll.

Courtesy of Monroe Energy.

Courtesy of Michael S. Hildebrand.

Courtesy of Michael S. Hildebrand.

**FIGURE 5-14** A, B, C, and D Foam concentrate can be stored in tankers, tanks, totes, drums, or cans. The manufacturer's data sheet should be consulted for application rates.

*NOTE:* The formula (0.785)(diameter$^2$) is commonly used during the preincident planning process, while the formula (0.8)(diameter$^2$) is commonly used for field applications to simplify the calculation.

- Calculate Dike or Rectangular Area Around the Tank Area = Length × Width

3. *Determine the recommended NFPA 11 foam application rate, as noted above.*
4. *Determine the duration of foam application per NFPA 11.*
   - Flammable liquid spill = 15 minutes
   - Storage Tank
     - Flash Point 100°–200°F = 50 minutes
     - Flash Point < 100°F = 65 minutes
     - Crude Oil = 65 minutes
     - Polar Solvents = 65 minutes
     - Seal Application = 20 minutes
5. *Determine the quantity of foam concentrate required. This figure will be determined by the percentage of foam concentrate used.*

Successful extinguishment of a storage tank fire will require sufficient firefighting foam at the recommended application rate and for the prescribed application duration. To accomplish this goal, emergency responders will require not only adequate foam and water supplies but also the capability to proportion, move, and effectively apply the firefighting foam onto the hazard. Obviously, the IC who has done some homework before the incident and predetermined the foam concentrate and water flow requirements for each tank in a facility will be further along the incident timeline than the IC who "wings it" on gameday with a calculator. See Scan Sheet 5-C for an example of how to do foam calculations.

# Scan Sheet 5-C

**PROBLEM 1:** A 125-ft.-diameter open-top floating roof tank containing gasoline is fully involved in fire. Determine the amount of foam concentrate required to control and extinguish the fire. The fire department is using a 3% × 3% alcohol-resistant foam concentrate (ARC).

© Jones & Bartlett Learning.

1. ***What is burning?***

   Gasoline–hydrocarbon liquid. The 3% × 3% ARC can be used and will be proportioned at a 3% concentration.

2. ***Determine the surface area involved.***

   Area = (0.8)(diameter$^2$)
   Area = (0.8)(125 ft.)$^2$
   Area = 12,500 ft$^2$

3. ***Determine the appropriate foam application rate.***

   The fire department is using portable application devices—foam cannons.

   The foam application rate is 0.16 gpm/ft$^2$

   Foam application = area × recommended application rate
   Foam application = (12,500 ft$^2$)(0.16 gpm/ft$^2$)
   Foam application = 2,000 gpm

4. ***Determine the duration of foam application.***

   Gasoline has a flash point of approximately −45°F. Therefore, the recommended duration is 65 minutes.
   Required foam solution = foam application × duration
   Required foam solution = (2,000 gpm)(65 minutes)
   Required foam solution = 130,000 gallons

5. ***Determine the quantity of foam concentrate required.***

   The fire department is using a 3% foam concentrate.
   Required amount foam concentrate = required foam solution × 3%
   Required amount foam concentrate = (130,000 gallons)(0.03)
   Required amount foam concentrate = 3,900 gallons
   Required amount of water = 126,100 gallons

**PROBLEM 2:** A 150-ft-diameter covered floating roof tank containing gasoline has overflowed and ignited. Both the tank and the dike area (100 ft. × 80 ft.) are completely involved in fire. Determine the amount of foam concentrate required to control and extinguish the fire. The fire department is using a 3% × 3% alcohol-resistant foam concentrate (ARC).

1. ***What is burning?***

   Gasoline–hydrocarbon liquid. The 3% × 3% ARC can be used and will be proportioned at a 3% concentration.

2. ***Determine the surface area involved.***

| *Storage Tank* | *Dike Area* |
|---|---|
| Area = (0.8)(diameter$^2$) | Area = (length)(width) |
| Area = (0.8)(150 ft)$^2$ | Area = (100 ft.)(80 ft.) |
| Area = 18,000 ft$^2$ | Area = 8,000 ft$^2$ |

3. ***Determine the appropriate foam application rate.***

   The storage tank is protected with foam chambers designed for full-surface protection and requires a foam application rate of 0.10 gpm/ft$^2$. Portable application devices are required for the dike fire and require a foam application rate of 0.16 gpm/ft$^2$.

   *Storage Tank*
   Foam application = area × recommended application rate
   Foam application = (18,000 ft$^2$)(0.10 gpm/ft$^2$)
   Foam application = 1,800 gpm

   *Dike Area*
   Foam application = area × recommended application rate
   Foam application = (8,000 ft$^2$)(0.16 gpm/ft$^2$)
   Foam application = 1,280 gpm

4. ***Determine the duration of foam application.***

   Gasoline has a flash point of approximately −45°F. Therefore, the recommended duration is 15 minutes for the dike area and 65 minutes for the storage tank.

   *Storage Tank*
   Required foam solution = foam application × duration
   Required foam solution = (1,800 gpm)(65 minutes)
   Required foam solution = 117,000 gallons

   *Dike Area*
   Required foam solution = foam application × duration
   Required foam solution = (1,280 gpm)(15 minutes)
   Required foam solution = 19,200 gallons

5. ***Determine the total quantity of foam concentrate required.***

   The fire department is using a 3% foam concentrate.

| *Storage Tank* | |
|---|---|
| Required amount foam concentrate | = required foam solution × 3% |
| Required amount foam concentrate | = (117,000 gallons)(0.03) |
| Required amount foam concentrate | = 3,510 gallons |
| Required amount of water | = 113,490 gallons |
| *Dike Area* | |
| Required amount foam concentrate | = required foam solution × 3% |
| Required amount foam concentrate | = (19,200 gallons)(0.03) |
| Required amount foam concentrate | = 576 gallons |
| Required amount of water | = 125,524 gallons |
| Total amount foam concentrate required | = 4,086 gallons |
| Total amount of water required | = 239,014 gallons |

## Firewater Supply and Delivery Systems

Determining foam concentrate requirements is only half of the equation for determining storage tank firefighting capability; a sustainable water supply is needed to provide cooling water and to make foam solution. Large diameter aboveground storage tank fires require big water supplies, especially when more than one tank is burning. The December 11, 2005, fire in Buncefield, England, required 14.5 million gallons (metric) of water! See Scan Sheet 5-D for additional information on the case study.

It is important for firefighters to know the types of firewater systems used for the protection of petroleum distribution facilities and their capabilities. When evaluating firewater systems in tank farms, terminals, and bulk storage plants, what is on paper (e.g., preincident plans, firewater tests) is sometimes not consistent with the actual capabilities and performance of the firewater system. (See FIGURE 5-15.)

Water supplies for bulk storage tank firefighting consist of two major components: the water supply source and the water supply delivery system. Understanding each component's capabilities is critical to the success of the firefighting operation. See Scan Sheet 5-D.

- **Water Supply Source.** Where will the firewater come from? Is there sufficient quantity to support long-term firefighting operations with large-volume fire flows?
- **Water Supply Delivery System.** How will the water be moved from its source to where you will need it? In evaluating firewater systems at fixed facilities, what are the number, size, and power source of the fire pumps? What is the flow capacity of the firewater system (i.e., gallons per minute)? What are the static pressure (water at rest in the fire water main) and the residual pressure (water in motion as it leaves the hydrant)? Will there by adequate residual pressure in the water system when multiple delivery devices are operating?

FIGURE 5-15 Firefighters should be very familiar with the facility water system, its capabilities, and weaknesses before an incident. Planning, training, and exercises are the key to a successful response.

Courtesy of William T. Hand.

# Scan Sheet 5-D—Case Study: The Buncefield Incident At Hemel Hempstead, England

Early on Sunday, December 11, 2005, a series of explosions and subsequent fire destroyed large parts of the Buncefield oil storage and transfer depot located at Hemel Hempstead, England. The main explosion took place at 06:01 hours. It was massive—causing widespread damage to neighboring properties. This was followed by a large fire that eventually engulfed 23 large fuel storage tanks. The incident injured 43 people and caused significant damage to commercial and residential properties near the Buncefield site. Approximately 2,000 people were evacuated from their homes. Sections of the M1 highway were closed. Some schools in Hertfordshire, Buckinghamshire, and Bedfordshire were closed for 2 days following the explosion. The fire burned for 5 days, destroying most of the site and emitting a large plume of smoke into the atmosphere that dispersed over southern England and beyond.

© Chiltern Air Support Unit/Hertfordshire Police/PA Wire/AP Photos.

© Hertfordshire Police.

The incident scenario started late on Saturday, December 10, when a delivery of gasoline started to arrive by pipeline at Tank 912. The safety systems in place to shut off the supply of gasoline to the tank to prevent overfilling failed to operate, resulting in gasoline flowing down the side of the tank and collecting at first inside the diked area. As the overfill continued, a vapor cloud formed by the mixture of gasoline and air flowed over the dike wall, dispersed, and flowed west offsite toward the Maylands Industrial Complex. A white mist was clearly observable in closed-circuit TV footage produced during the investigation. The exact nature of the mist is not known but it is believed to have been a volatile fraction of the fuel (butane) or ice particles formed from the chilled, humid air as a consequence of the evaporation of the escaping fuel.

An estimated 300 tons (approximately 96,000 gallons) of gasoline escaped from the tank, about 10% of which turned to vapor that mixed with the cold air and eventually reached flammable concentrations. The main explosion at Buncefield was unusual because it generated much higher overpressures than would usually have been expected from a vapor cloud explosion. The exceptional scale of the incident was matched by the scale of the emergency response. A "Gold Command" (the U.K. equivalent of a U.S. NIMS Type I incident management team) was established within hours and coordinated by Hertfordshire Police. Unified command included the Hertfordshire Fire and Rescue Service, Hertfordshire County Council, Dacorum Borough Council, and the Environment Agency with Health and Safety Executive in support.

At the peak of the fire, at noon on Monday (now December 12), 25 Hertfordshire fire apparatuses were onsite with 20 support vehicles and 180 firefighters. The full-scale operation involved 1,000 firefighters from Hertfordshire and across the country, supported by police officers from throughout the UK. It took 32 hours to extinguish the main blaze, although some of the smaller tanks were still burning on the morning of Tuesday, December 13. The following day a new fire started in a previously undamaged tank, but the fire service let it burn out safely. Firefighting efforts required 198,129 gallons (750,000 liters) of foam concentrate and 14,529,462 gallons (55 million liters) of water.

*Sources:* The following sources were referenced for this case study:

Major Incident Investigation Board, "*The Buncefield Incident 11 December 2005, Volume 1*," Health and Safety Executive: Kew, Richmond, Surrey, England (2008).

Major Incident Investigation Board, "*The Buncefield Incident 11 December 2005, Volume 2b: Recommendations on the Emergency Preparedness For, Response to and Recovery from Incidents*," Health and Safety Executive: Kew, Richmond, Surrey, England (July 2007).

## Water Supply Sources

Storage tank fires often tax primary water supplies to the point that alternate sources will be required as the incident progresses. It is important to investigate and evaluate alternative water supply sources, incorporate them into preincident plans, and conduct regular drills and exercises so that all the "bugs" and surprises are worked out ahead of time.

If the water supply source is not located close to the potential hazard, fire pumps, storage tanks, and long hose lays will be required. The reliability of a water supply system decreases as the complexity of the system increases. The final preincident plan should go with the simplest delivery system. Common water supply sources include rivers, reservoirs, wells, elevated tanks, and ground storage facilities.

### Rivers and Reservoirs

Natural water supplies such as lakes, streams, and rivers are common firewater sources. Large public and private reservoirs are also frequently used to supply domestic water. The big advantage of these sources is that they are practically unlimited water supplies. However, experience has proven they can be vulnerable to icing, environmental contamination, debris that plugs fire pump intakes, or droughts or unusually low tide water levels that drop water levels below pump suction intakes.

### Fire Water Retention Ponds

Retention ponds can be found at many facilities and can be a supplemental source of firewater. These ponds are typically rubber lined and are used to collect normal rain runoff from throughout the facility. In addition, the pond can be filled from a source like a water main. (See **FIGURE 5-16**.)

**FIGURE 5-16** Firewater retention ponds can be an effective supplemental source of firewater.

Courtesy of Gregory G. Noll.

### Wells

Wells are sometimes used to provide supplemental water supplies. However, they can be adversely affected by seasonal groundwater levels and local consumption rates. Wells are sometimes used to fill storage facilities such as elevated tanks. Pumps are used to move the water into the tank. If the primary water supply is from a gravity feed tank supplied by a well, anticipate having serious water supply problems with any storage tank fire. Most well pumps that supply the water storage tank cannot fill the tank faster than the water that is being removed to fight the fire.

### Elevated Tanks

Elevated tanks provide a passive capability to supply firewater. Tank water is supplied to the ground by gravity. Most elevated tanks are 110 to 130 feet high; therefore, water pressures are approximately 50 to 65 psi (metric)—not very impressive but helpful.

### Ground Storage Facilities

Ground level water storage usually consists of ground level welded steel tanks or embankment supported rubberized fabric tanks also known as ERSF tanks. These are cost-effective, large-capacity firewater systems, and multiple water supply locations may be found at large facilities.

## Fixed Pumping and Delivery Systems

Petroleum manufacturing facilities, such as refineries, typically have a fixed pumping system to support the firewater system. In contrast, petroleum distribution facilities often rely upon the municipal water system to support any firefighting operations. Facilities with an onsite firewater system will have pumps to boost the firewater system pressure and help move the water from the storage source, through firewater mains, and eventually to fire hydrants located throughout the facility.

Fire pumps are like the family dog—they need to be loved, fed, and exercised on a regular basis in order to stay happy and healthy. Like any mechanical piece of equipment, fire pumps can fail when they are needed the most because of poor maintenance or power failures. Experience shows that no single fire pump should be relied upon as the sole source for supplying firewater. A well-engineered facility usually has two or more fire pumps, each having the ability to provide the required fire flow, or a series of smaller pumps, any one of which can supply half of the demand. Both of these approaches should provide sufficient redundancy if one fire pump should fail. Many facilities also use alternative options to supply power to the fire pumps (e.g., steam, electrical, gasoline/diesel–driven fire pumps). That way, a power failure involving one source of energy will not disable the entire firewater system. (See FIGURE 5-17.)

### Centrifugal Fire Pumps

Centrifugal pumps are the most common type of fire pump found at industrial facilities. The centrifugal pump is compact, very reliable, and easy to maintain. This pump can be driven by electric motor, steam turbine, or gasoline and diesel engines. Centrifugal pumps are available in horizontal

FIGURE 5-17 Fire pumps and their drivers (power source) are a critical element of the firewater system and need to be inspected, maintained, and exercised on a regular maintenance schedule.

Courtesy of Michael S. Hildebrand.

or vertical designs and have capacities to approximately 4,500 gpm (17,034 g/L). Centrifugal fire pumps are rated at pressures from 75 to 280 psi (5.7 to 19.3 bar).

### Vertical Turbine Pumps

Vertical turbine pumps are the most common fire pump used to supply water from lakes, rivers, and reservoirs. They are often used where water can be taken from an atmospheric water source and when the draft is less than 50 ft (15.2 m).

## ■ Portable Pumping and Delivery Systems

### Trailer-Mounted Horizontal Pumps

These are basically large fire pumps permanently mounted on a trailer. Their main advantage is mobility and the ability to support fixed firewater systems. Several manufacturers offer high-capacity, trailer-mounted pumps in the range of 4,400 to 7,500 gpm (16,600 to 28,390 L). Portable pumps have proven to be versatile in industrial mutual aid groups where many members can bring several trailer-mounted pumps to the scene with trailer-mounted, high-capacity monitors. With planning, training, and exercises, fire flows of 12,000 gpm (45,424 g/L) can be established in a reasonable amount of time using large diameter hose and manifolds. In addition to firefighting, these pumps can also be used as an emergency water supply for manufacturing purposes, or dewatering flooded areas. (See FIGURE 5-18.)

### Trailer-Mounted Submersible Pumps

Many refineries and storage tank facilities are located along waterways that are tidal. During an unusually low tide, water intakes for fixed pumping systems may be unable to take suction. Submersible lift pumps can overcome this problem. The Williams DEPENDAPOWER™ trailer-mounted hydraulic submersible diesel-driven pump can pump water at a rate of 8,000 gpm (30,000 L) from a water source as far as 200 ft (61 m) away with a vertical lift of 35 ft (10.6 m). This water flow can then be delivered to an onboard or separate boost pump where pressure is increased before sending to end-of-line devices. (See FIGURE 5-19.)

Courtesy of Gregory G. Noll.

Courtesy of Gregory G. Noll.

Courtesy of Michael S. Hildebrand.

Courtesy of Michael S. Hildebrand.

FIGURE 5-18 Trailer-mounted pumps provide mobility, flexibility, and the ability support fixed firewater systems.

### Mobile Fire Apparatuses

As noted earlier, municipal, industrial, or military fire apparatuses are flexible in their ability to pump water from hydrants or by draft. Industrial fire pumpers have the ability to deliver large volumes of water.

**FIGURE 5-19** Trailer-mounted submersible pumps can be used to draft water from any stationary water source.

Courtesy of Monroe Energy.

## Determining Water Supply Requirements

Storage tank fires can require tremendous quantities of water for a sustained period of time. (See **FIGURE 5-20**.)

Before an effective fire attack can be made, it must be determined if the water system is capable of:

- Delivering a water flow rate equal to or greater than that required to control the largest potential fire area
- Delivering the required flow rates at pressures that can be effectively used by water application devices (e.g., portable monitors, handlines)

There are several assumptions regarding water supply requirements that are often misunderstood by emergency response personnel, especially when relying upon petroleum facility water supplies. These include:

1. **Assumption:** The fire water system is known to be adequate for fire attack because all of the past fires at the facility have been successfully controlled.

   **Reality:** Most fires never reach their full potential because they are extinguished in their incipient stages. Therefore, they were controlled well before they reached their full-scale potential and never really taxed the water system.

2. **Assumption:** The water flow rate available in the facility is believed to be adequate because the rated flow of the facility's fixed fire pumps exceeds the maximum foreseeable water flow demand.

   **Reality**: Fire pumps are not constant flow and constant pressure devices. Therefore, the sum of their rated flows is not equal to the water flow rate available in the facility. When delivering water, fire pumps must provide sufficient pressure to operate water application devices in the area of the fire and to overcome pressure losses in the piping system. If this pressure is higher than the pumps' rated

**FIGURE 5-20** A sustainable water and foam concentrate supply are a critical role in establishing and sustaining the necessary water and foam solution flows at a storage tank emergency.

(left) Courtesy of Monroe Energy; (right) Courtesy of William T. Hand.

pressure, they will flow LESS water than the sum of their flows. If the water flow rate demand exceeds the rated flow of the pumps, the pressure available in the fire area will decrease, with possible effects on the operation of water application devices.

As a result, the water flow rate available in each area of a facility varies depending upon their relative elevation and location with respect to the pumps and the pressure requirements of the water application devices to be used. In addition, experience also shows that some water systems have been modified or damaged over time. While they may have been properly designed and installed, years of neglect may have rendered them ineffective.

3. **Assumption:** The normal/static pressure is an adequate indicator of the firewater system's capability (e.g., the system is adequate because it has a pressure of 100 psi [6.8 bar] when no water is flowing).

   **Reality:** As soon as water begins to flow, the operating pressure will drop. The greatest pressure drop will be in the area where the water is being used. How far the pressure drops will be dependent on the pressure available at the system's water source(s), the distance between the fire and the system's water sources, the size and arrangement of the piping in the firewater system, and the water flow rate. Remember, the greater the flow rate, the greater the pressure drop.

See Scan Sheet 5-E for an example of how these three problems can work against emergency responders.

# Scan Sheet 5-E—How a Firewater System Reacts to the Demands of a Flammable Liquids Storage Tank Firefighting Operation

**Point A**: During the early stages of a fire, before a fire attack on the storage tank or dike fire is made, several monitors, master streams, or handlines may be placed in service to protect exposures. As a result, the available water pressure in the immediate fire area drops slightly.

**Points B and C:** When the fire attack on the burning tank or dike is initiated, additional water streams are ordered into service. As these additional master streams, monitors, or hoselines and devices are placed into service, the water pressure continues to drop even further. At this point one of two outcomes is possible:

1. The flow required to control the fire is obtained and the fire is brought under control, with the flow and water pressure remaining fairly constant, or (2) the water pressure in the area begins to fall below that required to maintain full firefighting effectiveness.
2. The overall water flow rate cannot be significantly increased because placing additional devices in service will further lower available water pressure, causing the flow out of every line or device already in operation to decrease. As a result, the fire continues to grow, possibly out of control or spreading to adjacent exposures, while the water flow rate remains fairly constant. This is characteristic of an inadequate water system, as illustrated in the chart below.

Reprinted with permission by Loss Control Associates, Inc.©

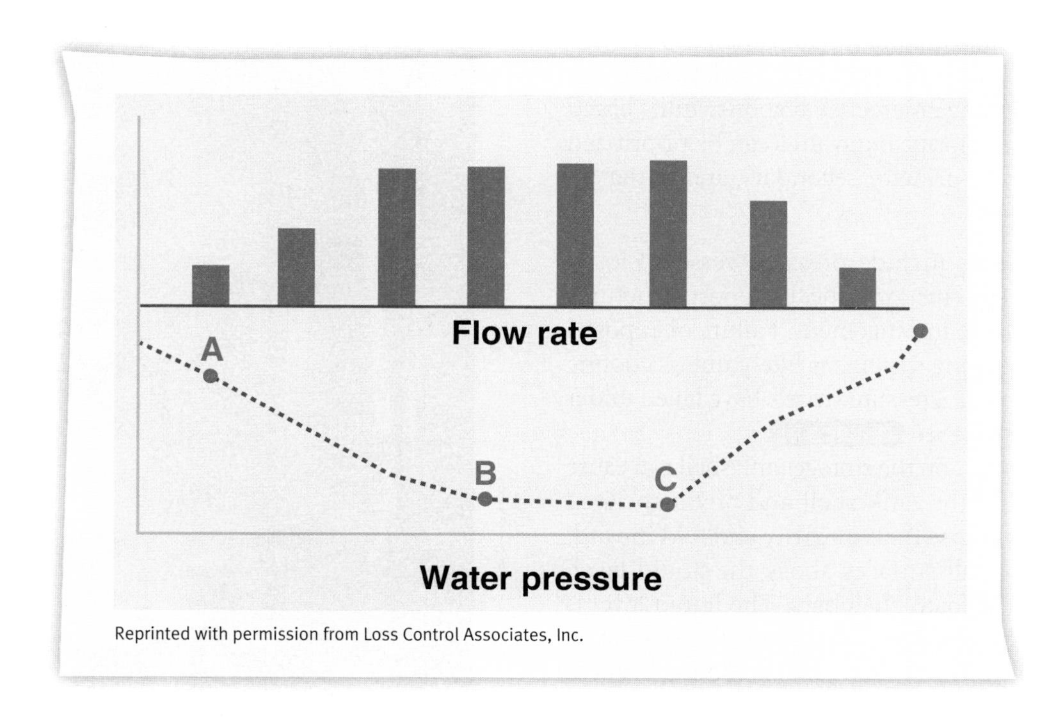


Reprinted with permission from Loss Control Associates, Inc.

## Determining Cooling Water Requirements for Exposures

Cooling water required for exposure protection taxes the firewater system and can compete with both fire streams and fixed suppression systems dedicated to fire extinguishment. Excessive cooling water can also lead to runoff problems where floating hydrocarbon contaminants create safety hazards, overflow dikes, and significantly stress the facility stormwater and wastewater pumps and separators (e.g., API Separator). Loss of power and utilities to facility wastewater pumps may also cause runoff to back up into the fire area and overflow dikes and other tank containment measures.

Refineries and bulk storage terminals can have extensive water runoff control systems designed to channel rainwater to water pollution collection systems. If the wastewater separator fails due to flooding, polluted water will flow to the facility outfall, resulting in serious pollution problems and creating another incident. Much like a ship on fire, water in must equal water out or the ship sinks. Water runoff should be addressed early in the fire by facility engineering and wastewater specialists.

Water applications for exposure protection usually start before foam application is initiated. Exposure lines should be applied when there is direct flame impingement on exposed tanks or when radiated heat is sufficient to cause steam at the tank shell when water is applied. Exposure streams should be shut down when steam is no longer produced from the metal surface. Continuously applying cooling streams onto exposures can take water away from fire suppression operations, as well as create additional water runoff problems.

Cooling water can be conserved by prioritizing fire exposures and assigning emergency response units based upon the exposure's priority. Exposures can be prioritized into three basic levels: primary, secondary, and tertiary.

### Primary Exposures

Primary fire exposures include pressure vessels, closed containers, piping systems, or critical support structures exposed to direct flame impingement. Failure of exposed vessels, tanks, and piping systems is likely unless cooling water is quickly applied. Pressure vessels have failed under 10 minutes into a fire. (See **FIGURE 5-21**.)

Direct flame contact on the storage tank shell can cause the upper portion of the tank shell and any associated fixed foam systems to lose their integrity and fold inward. Streams should cool all surfaces above the liquid level, especially around the foam chambers. The liquid level is usually indicated by the discoloration on the tank shell due to the paint burning off. Remember, cooling water is a valuable resource—don't waste it!

Be careful of applying water onto adjacent open floating roof tanks, because sinking the floating roof with cooling water streams is expensive. Applying cooling water to structural steel and pressure vessels in the first 10 minutes is essential. It may also result in an ignition, thus creating another incident. It also won't look good on your resume.

### Secondary Exposures

Secondary fire exposures include pressure vessels further away from primary exposures, and closed containers, piping systems, or critical support structures exposed to radiant heat. Failure of structural components is possible if cooling water is not applied.

### Tertiary Exposures

Tertiary or third-level fire exposures include noncritical exposures without life safety concerns. These could include low value buildings that have been evacuated, such as storage buildings and sheds. Remember, radiant heat will pass through structures with clear glass and windows. In addition to applying exterior cooling lines, firefighters should be sent inside to check for any fire extension.

The actual amount of cooling water required on the fireground will be dependent upon the type and area of

**FIGURE 5-21**

Refinery Terminal Fire Company.

exposure. Basic cooling water guidelines for *exposed* tanks and pressure vessels are as follows:

- Atmospheric storage tanks up to 100 ft. diameter require 500 gpm.
- Atmospheric storage tanks from 100 to 150 ft. diameter require 1,000 gpm.
- Atmospheric storage tanks exceeding 150 ft. diameter require 2,000 gpm.
- Pressure vessels should have a *minimum* of 500 gpm applied at the point of fire impingement. This is a widely quoted number which has proven to be a reliable guideline over time. Taking action with less water using offensive and defensive tactics increases risk to personnel significantly. However, lower flow rates may still be effective if they are applied from fixed systems and can be activated without risk to emergency responders.

## Summary

There are two basic categories of fuels. These include standard hydrocarbon fuels and polar solvent fuels.

Class B firefighting foam remains the workhorse of storage tank firefighting. Chemically, Class B foams can be divided into two general categories: synthetic based or protein based.

Firefighting foam concentrates should be selected based upon the type of fuel (e.g., hydrocarbon vs. polar solvent) and the type and nature of the hazard to be protected (e.g., spill scenario vs. storage tank). The most common foam concentrates used for flammable liquid storage tank fire protection are fluoroprotein foam, aqueous film forming foam (AFFF), alcohol-resistant AFFF (ARC), and film forming fluoroprotein (FFFP) foam.

Bulk flammable liquid storage facilities that are designed to meet the requirements of NFPA 30—*Flammable and Combustible Liquids Code* and American Petroleum Institute (API) standards have many safety and fire protection features engineered into them that can favorably influence the behavior of the products in the event of a spill or fire. These systems and their design features should be factored into the analysis during preincident planning and response operations. Understanding how these systems function and their limitations could mean the difference between a good and bad outcome.

When storage tanks are not protected by fixed and semifixed systems, emergency responders must then rely upon manual firefighting efforts to control and extinguish the fire. This may include portable foam monitors and handlines, mobile fire apparatuses, and trailer-mounted pumps and monitors.

The availability of water and foam concentrate are critical factors in evaluating the risks involved in a storage tank emergency. If an adequate water and foam supply is not available for both protecting exposures and controlling the fire, the IC should consider implementing defensive or nonintervention tactics until sufficient resources are available. As a general tactical guideline, foam application operations should not be initiated until sufficient foam concentrate is onsite to extinguish 100% of the exposed flammable liquid surface area.

Successful extinguishment of a storage tank fire will require sufficient firefighting foam at the recommended application rate and for the prescribed application duration. To accomplish this goal, emergency responders will require not only adequate foam and water supplies but also the capability to proportion, move, and effectively apply the firefighting foam onto the hazard.

## References and Suggested Readings


1. *A Firefighter's Guide to Foam*, Exton, PA: National Foam (2002).
2. *General Foam Information, Data Sheet # D10D03010*, Mansfield, TX: Chemguard Specialty Chemicals and Equipment (2005).
3. International Fire Service Training Association, *Principles of Foam Firefighting* (2nd edition), Stillwater, OK: IFSTA (2003).
4. National Fire Protection Association, *NFPA 11 – Standard for Low, Medium, and High Expansion Foam* (2016 edition). Quincy, MA: National Fire Protection Association (2016).
5. National Fire Protection Association, *NFPA 30 - Flammable and Combustible Liquids Code* (2016 edition). Quincy, MA: National Fire Protection Association (2016).
6. Noll, Gregory G., and Michael S. Hildebrand, *Hazardous Materials: Managing the Incident* (4th edition, pp. 352–365). Burlington, MA: Jones and Bartlett Learning (2014).

# CHAPTER 6

# Tactical Response Guidelines

Courtesy of Gregory G. Noll.

## Objectives

1. Describe the following hazards, safety procedures, and tactical guidelines for managing the following storage tank safety problems and issues, including:
   - Confined spaces
   - Hydrogen sulfide
   - Boilover
2. Describe the causes, hazards, and methods of handling the following conditions as they relate to fires involving flammable liquid storage tanks (16.2.2.4):
   - Frothover
   - Slopover
   - Boilover
3. Given a flammable liquid bulk storage tank involved in a fire, identify the factors to be evaluated as part of the risk assessment process, including (16.2.2.2):
   - Type of storage tank
   - Product involved
   - Amount of product within the storage tank
   - Nature of the incident (e.g., seal fire, tank overfill, full-surface fire)
   - Tank spacing and exposures
   - Fixed or semifixed fire protection systems present
4. Given a flammable liquid storage tank fire, describe the methods and associated safety considerations for extinguishing the following types of fires using fixed systems or portable application devices (16.3.7):
   - Pressure vent fire
   - Seal fire on an open floating roof tank
   - Seal fire on an internal floating roof tank
   - Full surface fire on an internal floating roof tank
   - Full surface fire on an external floating roof tank
   - Dike fire
5. Describe the hazards, safety procedures, and tactical guidelines for handling a flammable liquid bulk storage tank with a sunken floating roof (16.3.6).
6. Describe the hazards, safety procedures, and tactical guidelines for handling the following types of emergencies involving low pressure horizontal and vertical storage tanks (16.3.7[1] and 16.2.2.2[4]):
   - Vent fire
   - Tank overfill
7. Describe the hazards, safety procedures, and tactical guidelines for handling product/water drainage and runoff problems that may be created at a flammable liquid bulk storage tank fire (16.3.9[6]).

## Chapter Outline


- Objectives
- Key Terms
- General Response Guidelines
- Storage Tank Safety Issues
- Boilover, Slopover, and Frothover
- Ground and Dike Fires
- Full-Surface Tank Fires
- Cone Roof Tank Fires
- Open Top Floating Roof Tank Fires
- Covered Floating Rook Tank Fires
- Horizontal and Vertical Low Pressure Tank Fires
- Summary
- References


## Key Terms

Boilover A violent ejection of flammable liquid from its container caused by the vaporization of water beneath the body of liquid. It may occur after a lengthy burning period of products such as crude oil when the heat wave has passed down through the liquid and reaches the water layer at the bottom of a storage tank. It will not occur to any significant extent with water-soluble liquids or light hydrocarbon products such as gasoline. Boilovers typically occur with liquids with a very wide range of boiling points, with crude oil being the most common.

Confined Space A space that (1) is large enough and so configured that an employee can bodily enter and perform assigned work, (2) has limited or restricted means for entry or exit (e.g., tanks, vessels, silos, storage bins, hoppers, vaults, and pits are spaces that may have limited means of entry), and (3) is not designed for continuous employee occupancy.

Flammable Atmospheres The flammable range for most common hydrocarbon liquids, such as gasoline, heating oils, diesel fuels, and aviation gasoline, typically ranges from 1% to 10% vapor-to-air mixture. However, the flammable range for oxygenated hydrocarbons such as alcohols and glycols are wider.

Frothover Can occur when water already present inside a tank comes in contact with a hot viscous oil being loaded. A typical example occurs when hot asphalt is loaded into a tank car that contains some water. The asphalt is initially cooled by contact with the cold metal, and there is no reaction. However, the water can later become superheated, eventually start to boil, and cause the asphalt to overflow the tank car.

Slopover Can result when a water stream is applied to the hot surface of a burning oil, providing the oil is viscous and its temperature exceeds the boiling point of water. It can also occur when the heat wave contacts a small amount of stratified water within a crude oil. As with a boilover, when the heat wave contacts the water, the water converts to steam and causes the product to "slopover" the top of the tank. Slopovers can range from a quietlike boiling of the product over the tank to a large explosion of burning slop.

## General Response Guidelines

### Storage Tank Fire Experience

While aboveground storage tank emergencies are low probability events, they often have high consequences.

A 2008 study conducted by the National Fire Protection Association (NFPA) of storage tank fires using data from the U.S. Fire Administration's National Fire Incident Reporting System (NFIRS) examined the number of incidents, location, cause, etc. The study revealed that from 2007 through 2011 there were an estimated average of 301 storage tank fires per year that required a response from a municipal fire department. Note that NFIRS does not capture responses by industrial fire departments or fire brigades; therefore, we can assume that the actual number of responses to storage tank fires is higher than NFIRS's data (Campbell, 2008).

A study published in 2006 in the *Journal of Loss Prevention in the Process Industries* reported on 242 storage tank incidents spanning more than 40 years (1960 to 2003) showed that 74% of storage tank accidents occurred in petroleum refineries, oil terminals, or storage facilities. Of these 242 accidents, 80 incidents (33%) were caused by lightning strikes and 73 incidents (30%) were caused by human error (e.g., tank overfill). Other causes included equipment failure, sabotage, cracking and rupture of the tank shell, leak and line rupture, static electricity, and open fames (e.g., cutting and welding). Of these 242 incidents, 114 occurred in North America, 72 in Asia, and 38 in Europe. The most frequently involved locations were bulk storage terminals and pumping stations (Chang & Cheng-Chung, 2006).

### General Tactical Guidelines

This chapter will provide general guidelines for handling common types of storage tank emergencies. We will identify the common types of emergencies associated with each type of storage tank and summarize the tactical priorities for handling the associated problems.

The following basic principles can be applied to most tank firefighting situations:

1. Storage tank emergencies can present the incident commander with multiple tactical problems that must be sorted out and prioritized. Consider the following general risk-based guidelines:
   - If the situation involves both a full surface tank fire and the surrounding dike area, the dike area may be extinguished before attempting to extinguish the full surface fire. Nonetheless, there have been successful operations where both the dike area and storage tank have been simultaneously and successfully controlled.
   - If the situation involves two storage tank fires, where tank 1 is a full-surface fire and tank 2 is a seal fire on an open floating roof tank, tank 2 should be extinguished first.
   - If the situation involves two simultaneous storage tank fires, the IC must evaluate the following risk factors, including:
     - Size and type of storage tanks involved
     - Product involved and the level of product within each tank
     - Installation of fixed or semifixed foam systems
     - Ability to extinguish the fires (e.g., foam concentrate, water, and application device requirements)
2. If a fire occurs while product is being transferred into or out of a tank, stop the product flow until the situation can be fully assessed. For example, removing product from a tank involved in a full

surface fire before fire control operations can be initiated may lead to severe tank shell damage above the product line, causing the tank shell to fold in.

3. Think big early and often. Adequate foam supplies must be on hand before any foam attack operations can be initiated, and the water supply system must be capable of sustaining the foam operation. Based upon NFPA 11 recommendations, a flammable liquid spill fire will require approximately a 20-minute duration of application, while a flammable liquid will require a 65-minute duration of application.
4. Foam and water logistics are critical to the success of firefighting operations. Offensive foam firefighting operations should only be initiated when all of the necessary resources are in place to meet the operational requirements of the incident. Resources would include foam concentrate and water supplies for at least 65 minutes of application, water pumping and movement, and foam proportioning and application devices. Likewise, the need for dewatering operations to ensure that dikes or other low areas do not overflow will be critical as response operations are continued.
5. The ability to extinguish a storage tank fire will be directly dependent upon the ability of emergency responders to apply sufficient foam *that actually reaches the fuel surface for the specified period of time*. Remember that the thermal updrafts and the winds created by the fire are the natural enemy of the foam. Just because 2,000 gpm (81,492 L/m) of foam is leaving the nozzle does not mean that 2,000 gpm (81,492 L/m) is actually reaching the fuel surface.
6. For a full-surface storage tank fire, emergency responders normally expect to see some progress within 20 to 30 minutes after initiating foaming operations. In addition, large storage tank fires are logistically demanding. The operations section cannot extinguish the fire unless there is timely and integrated support provided by the logistics section.
7. Always have a Plan B. Tank fires are dynamic events and Murphy was an optimist! The IC should continuously evaluate the situation and modify the incident action plan (IAP), as necessary. Typical problems that have occurred at previous tank fires have included changes in wind direction, failure of fire apparatus pumps or proportioners, and loss of water supplies. (See Scan Sheet 5-B in Chapter 5.)
8. Controlled burndown (CBD) may be an appropriate option when comparing the environmental impacts of offensive versus defensive strategies. Impacts may include the effects of particulates and environmental emissions, foam and firewater runoff, and groundwater and waterway pollution. For example, extensive air quality measurements taken during and after the Buncefield, England, fire by government environmental agencies did not identify any significant widespread air quality effects at the ground level.
9. Just because the fire goes out doesn't mean that the situation is necessarily over. Foam will still be required to control flammable vapor emissions. Depending upon the scenario, it may be necessary to secure the fuel surface of the tank for several days to prevent reignition. Site safety should be maintained until all of the product has been removed or transferred.

## Storage Tank Safety Issues

Before examining specific tactical problems and solutions, it's important to understand several safety issues that are common to many storage tank emergencies. These include the hazards and risks associated with confined spaces, the presence of hydrogen sulfide, and a phenomenon known as boilover.

### Confined Spaces

The OSHA Confined Space regulation (29 CFR 1910.146) defines a confined space as: "Any area that has limited or restricted means for entry or exit; is large enough and so configured that an employee can bodily enter and perform assigned work; and is not designed for continuous employee occupancy."

Both petroleum storage tanks and diked areas can be classified as confined spaces. Storage tanks and diked areas can pose three potential hazardous atmospheres: flammability, toxicity, and oxygen deficiency.

#### Flammable Atmospheres

The flammable range for most common hydrocarbon liquids, such as gasoline, heating oils, diesel fuels, and aviation gasoline, typically is from 1% to 10% vapor-to-air mixture. However, the flammable range for oxygenated hydrocarbons such as alcohols and glycols is wider.

OSHA and most national consensus standards consider a flammable atmosphere containing concentrations of 10% or less of the lower flammable limit (LFL) as acceptable for conducting rescue operations. Flammable concentrations of 10% to 20% of the LFL are considered hazardous and should not be entered by rescue teams unless they have proper PPE and respiratory protection and fire protection (e.g., backup hoseline), and all electric equipment is rated for Class 1, Division 2 atmospheres. As flammable limit readings rise above 20%, the level of risk to responders also rises. Rescue personnel operating in flammable concentrations of 50% of the LFL or greater should immediately leave the hazardous area.

#### Toxic Atmospheres

An atmosphere above the OSHA Permissible Exposure Limit (PEL) or the ACGIH Threshold Limit Value–Time

Weighted Average (TLV/TWA) does not necessarily prohibit entry into a confined space to perform rescue operations. However, the IC must understand that the risk to the rescue team also increases significantly and there is little or no margin for error. A rescuer who experiences damaged protective clothing or an air supply problem inside a confined space with a toxic atmosphere can face almost certain injury or death. Making an entry under these conditions is a risk-based decision that must be made on a case-by-case basis by both the IC and those actually assuming the risk.

### Oxygen-Deficient or Oxygen-Enriched Atmospheres

Oxygen-deficient confined spaces exist when the concentration of oxygen in air is less than 19.5%. The risk of entering an oxygen-deficient atmosphere is similar to entering a toxic atmosphere. Obviously, the lower the oxygen content, the greater the risk to the rescue team if there is an air supply problem.

OSHA defines an oxygen-enriched atmosphere as one where the concentration of oxygen in air is 23.5% or greater. Oxygen-enriched atmospheres also present rescuers with a significant risk as the increase in oxygen content increases the risk of fire. If a flammable atmosphere is present, the flammable range will also become wider.

When asked to cite examples of confined spaces, emergency responders often list areas such as underground vaults, sewers, or the interior of a storage tank. However, when OSHA's definition is applied to storage tank facilities, there are several special hazard areas that will require extra caution. These include tank dikes, ladders and walkways, and rooftops.

- **Tank Dikes.** Dikes are designed to contain flammable and combustible liquids in the event of an accidental spill or release, such as a tank overfill. The larger the tank capacity, the larger the diked area. As a general safe operating practice, diked areas should be treated as confined spaces during firefighting operations. Personnel entering diked areas should follow standard procedures for safe entry and accountability. Fire conditions can change rapidly in diked areas, and escape can be limited by the height of the dike, the slope of the dike wall, the dike's materials of construction, and obstructions such as piping. *NFPA 30—The Flammable and Combustible Liquid Code* has special provisions for access to and egress from a diked area where dike walls cannot exceed a height of 6 feet. (See **FIGURE 6-1**.)
- **Tank Ladders and Walkways.** Most storage tanks have ladders and walkways to traverse the tank dike or to gain access to the roof. Many tanks also have elevated walkways to access adjacent tanks without descending to the ground.

**FIGURE 6-1** Diked areas can be considered high-risk areas when conducting firefighting operations.

Courtesy of Jorge Carrasco.

**FIGURE 6-2** Storage tanks usually have only one way on and off the roof. Elevated walkways can be unsafe or dangerous during an emergency.

Courtesy of William T. Hand.

- Most large storage tanks do not have an alternate means of egress. If it is necessary to go to the roof using fixed walkways, it is important to plan an alternate escape route by placing fire department ground ladders or by positioning aerial ladders/platforms so that there is an alternate way off the roof. (See **FIGURE 6-2**.)

Tank walkways are especially hazardous under windy conditions, during periods of darkness, and when the walkways are covered with snow or ice. In addition, the wind girders on open floating roof tanks are typically not designed as a walking surface and may not have any rails.

Given that some seal fires have burned for days without creating significant tank problems, the IC must critically evaluate the need to place personnel on tank roofs. Personnel safety should not be compromised at the expense of operational expediency.

- **Tank Roofs.** Any storage tank roof could be a confined space, but open top and covered floating roof tanks present special hazards. As a general rule, emergency responders should treat all floating roof tanks as a hazardous area. *API Publication 2026—Safe Descent onto Floating Roofs of Tanks in Petroleum Service* considers any floating roof more than 8 feet below the top of the tank as a confined space. (See **FIGURE 6-3**.)

Medium and large diameter open-top storage tanks have ladders that are usually easy to negotiate; when the tank is in a nearly empty or full position, the ladder will always be in an awkward climbing position (e.g., nearly horizontal or vertical). On a covered floating roof tank, the ladder extends down from a manway on the fixed roof to the internal floating roof. Entry into this area is always a confined space. Even when a floating roof is in good condition and in a high position (nearly full), vapors may escape past the roof seals and gage pipe well seals. Vapors may also migrate through the pressure vacuum vent or accumulate in leaky pontoons. Always assume that the floating roof area is flammable. When crude oil is stored in the tank, hydrogen sulfide or sulfur dioxide (if tank is burning) may also be present.

The roof on an open-top floating roof tank is ordinarily constructed of steel, and mechanical or corrosion problems are usually visible from the top of the platform. However, when roof plates are in contact with the liquid, the condition of the underside of the roof is not visible and therefore a fall-through hazard is possible. Stay off the floating roof whenever possible; to minimize risks, handline operations for seal fires should be made from the gaugers platform or an aerial device.

**FIGURE 6-3** Open top floating roof tanks are considered confined spaces when the roof is 8 feet below the top of the tank. The ladder's angle adjusts as the roof goes up and down.

Courtesy of Tyler Bones.

## Hydrogen Sulfide

Hydrogen sulfide ($H_2S$) is a common hazard at petroleum storage facilities. $H_2S$ is a respiratory paralyzer, which upon entering the human body will "short circuit" the respiratory nervous system. Many petroleum industry employees and contractors have been killed over the years because of improper safe operating practices on or around storage tanks that contained petroleum liquids contaminated with $H_2S$. Failure to wear positive pressure air-supplied respiratory protection (e.g., SCBA, SAR) is a primary reason for death.

Emergency responders must take special precautions working around storage tanks that contain "sour" crude oils. The term "sour" refers to the presence of 0.5% and higher of hydrogen sulfide gas in a petroleum liquid, while "sweet" refers to the absence of $H_2S$.

$H_2S$ is a naturally occurring gas which results from the decomposition of metal sulfides and organic matter. Although we often look at it as a by-product of natural gas or oil drilling and production operations, it can also be encountered in sewers and septic tanks.

$H_2S$ is a flammable and poison gas which can paralyze the respiratory system, and can lead to unconsciousness, injury, or death. The most characteristic property is its strong odor, typically described as the smell of rotten eggs.

The threshold odor value is $<1.0$ ppm; however, smell cannot be relied upon to detect the presence of the gas as it deadens the sense of smell. The odor will seem to disappear in a few minutes, even though the gas is still present.

Current NIOSH exposure values for $H_2S$ are:

- TLV − TWA = 10 ppm
- TLV − STEL = 15 ppm
- IDLH = 300 ppm
- LEL = 40,000 ppm (4% concentration)

Typical effects of $H_2S$ exposures on individuals will include:

- 20 to 30 ppm = conjunctivitis
- 50 ppm = start to experience irritation of respiratory tract
- 150 ppm = objection to light; irritation of mucous membrane; headache
- 650 ppm = loss of consciousness
- 1,000 ppm = immediate acute poisoning
- 2,000 ppm = acute lethal poisoning

## Boilover, Slopover, and Frothover

### Boilover

Boilover is a phenomenon that is often misunderstood by storage tank owners, operators, and emergency responders. Boilovers have killed over 50 firefighters and injured hundreds of bystanders because they did not understand the hazards and risks or simply underestimated the area that would be effected by a boilover.

The NFPA Technical Committee on Flammable and Combustible Liquids defines a boilover as: "An event in the burning of certain oils in an open-top tank, when, after a long period of quiescent (causing no trouble or symptoms) burning, there is a sudden increase in fire intensity associated with expulsion of burning oil from the tank."

Boilover occurs when the residues from surface burning become more dense than the unburned oil and sink below the surface to form a hot layer that progresses downward much faster than the regression of the liquid surface. When this hot layer, called a "heat wave," reaches water or water-in-oil emulsion at the bottom of the tank, the water is superheated, then boils almost explosively, overflowing the tank.

Oils subject to boilover must have components with a wide range of boiling points, including light ends and a viscous residue. These characteristics are present in most crude oils and can be produced in synthetic mixtures.

For a boilover to occur, the following three conditions must be present:

1. *The tank must be involved in a full surface fire*. A floating roof seal fire will not produce a boilover as long as the roof is floating.
2. *The product must contain components with a wide range of boiling points, including light ends and heavy residues.* These characteristics are present in most crude oils and can also be produced in synthetic mixtures.
3. *The tank must contain a water bottom*. This water may be either free water or a water-in-oil emulsion, and normally occurs in tanks used to store crude oil. Water can also be introduced into the tank from firefighting operations.

Crude oils will burn at a rate of approximately 12 to 18 inches per hour. As the fire burns, the lighter hydrocarbons within the crude oil vaporize and are consumed by the fire. However, the heavier hydrocarbons (e.g., tars and asphalt) form a dense layer at temperatures of 300°F (149°C) and higher. This layer is often referred to as the "heat wave." The heat wave advances downward through the product at a rate of 1 to 4 feet (0.3 to 1.2 m) per hour. When the heat wave reaches any water in stratified layers within the crude oil or the tank's water bottom, the water will quickly expand to steam at a 1,700:1 expansion ratio and cause the tank's contents to be violently thrown from the tank. This action can create a fireball over 1,000 ft. (304.8 m) in diameter, and a burning wave of liquid that can travel over 500 ft. (152.4 m) at speeds up to 20 mph (32.18 kph). Evacuation of the immediate area should be considered as the heat wave approaches the bottom few feet of a crude oil tank.

The LASTFIRE (Large Atmospheric Storage Tank Fires) Project is a consortium of international oil companies that have reviewed the risks associated with storage tank fires and have developed industry best practices. In December 2016, the LASTFIRE Project published a boilover research position paper and lessons learned. Among the information noted for emergency responders was the following:

- Boilovers are an extreme fire event. It should be assumed that boilover will occur on a burning full surface fire crude oil tank if the fire is not extinguished in a relatively short period of time from ignition.
- There have been no documented cases of boilover on tanks where the fire was a rim seal fire only.
- While fire protection standards note "boilover," "slopover," and "frothover" as different events occurring due to different reasons, any scenario that involves the expulsion of hot or burning crude has the same potential for injury and damage.
- Boilovers can occur more than once on the same tank. Firefighters should not return to a tank, even if a boilover has occurred.
- Thermal imaging cameras or heat sensitive paint can help to assess the heat wave buildup, but cannot be totally relied upon as the heat zone buildup may not be uniform across the entire tank area.
- Apart from rapid extinguishment, none of the published theories to prevent or delay a boilover have been proven as practicable in real situations.
- Boilovers have occurred in fuels other than crude oil, where the liquid has been a mixture of products with a wide range of boiling points.

# Scan Sheet 6-A—Boilovers Have Killed and Injured Many Firefighters

### How Does a Boilover Occur?

After a long period of intense burning at the surface of the tank, the fire may appear to settle down and give the impression that it is burning slower and is actually safe to be in forward positions (e.g., operating master streams). (See Photo A.) After a period of quiescent burning which is marked by no significant burning activity, there is a sudden increase in fire intensity just before the expulsion of burning product from the tank. Responders have remarked that the fire at this point often produces a sizzling frying sound. The boilover occurs when the heat wave reaches freestanding water or water-in-oil emulsion in the bottom of the tank, converts the water to steam, and almost explosively overflows the tank. This burning liquid is projected up and over the dike area, then rains down on bystanders and emergency responders. (See Photo B.) It is estimated that a boilover can propel burning oil and vapor to a height 10 times the diameter of the tank. For example, photographs of a crude oil tank fire in 1926 showed a mass of flame estimated to be 1,100 ft. (335 m) in diameter and 6,000 ft. (1,829 m) high.

It is important to remember that a boilover can still occur even after extinguishment. The heat wave can continue to sink and come in contact with a water bottom.

### What Are the Indicators of a Potential Boilover?

The liquid level within the tank can be determined through the use of portable infrared sensors (i.e., thermal scanners) or hoselines. Thermal scanners are commonplace within fire departments and hazmat response teams. The primary advantage of thermal scanners is the ability of response personnel to stand back several hundred feet from the fire and safely determine the location of the heat wave and the cooler product.

Water streams can also be used to determine the heat wave and the product level. The tank wall above the liquid level will usually be somewhat distorted by the heat and fire. A water stream should be directed against the tank shell below the liquid level. If the water immediately vaporizes, it indicates that the heat wave is descending toward the bottom of the tank. The extent of the heat wave can then be assessed by locating the point at which the water stream no longer vaporizes. This delineates the bottom of the heat wave.

Reproduced courtesy of the American Petroleum Institute.

Reproduced courtesy of the American Petroleum Institute.

Reproduced courtesy of the American Petroleum Institute.

Reproduced courtesy of the American Petroleum Institute.

## ■ Slopover

A slopover can result when a water stream is applied to the hot surface of a burning oil, providing the oil is viscous and its temperature exceeds the boiling point of water. It can also occur when the heat wave contacts a small amount of stratified water within a crude oil. As with a boilover, when the heat wave contacts the water, the water converts to steam and causes the product to "slopover" the top of the tank. Slopovers can range from a quietlike boiling of the product over the tank to a large explosion of burning slop.

## ■ Frothover

A frothover can occur when water already present inside a tank comes in contact with a hot viscous oil which is being loaded. A typical example occurs when hot asphalt is loaded into a tank car that contains some water. The asphalt is initially cooled by contact with the cold metal, and there is no reaction. However, the water can later become superheated, eventually start to boil, and cause the asphalt to overflow the tank car.

A similar situation can arise when a tank used to store slops or residuum at temperatures below 200°F (93°C) and that contains a water bottom or oil-in-water (wet) emulsion receives a substantial addition of hot residuum at a temperature well above 212°F (100°C). After sufficient time has passed for the effect of the hot oil to reach the water in the tank, a prolonged boiling action can occur. This boiling action can rupture the tank roof and spread superheated froth over a wide area.

# Ground and Dike Fires

Ground fires and dike fires are the least severe tank farm scenario, and typically result from piping, valve, or pump leaks. In some instances, the incident may be due to another cause, such as operator error or an equipment malfunction. Incidents that result in a large release of product and the potential for vapor migration, such as a tank overfill scenario, have the potential to rapidly escalate.

## ■ Size-Up Considerations

Standard operating procedures (SOPs) should be used by the incident commander to ensure that site safety practices are implemented. If the tank has been preplanned, a copy of the preincident plan should be obtained. Initial questions and tasks that should be addressed as part of the hazard assessment and risk evaluation process include:

- What type of flammable liquid is involved (hydrocarbon or polar solvent)?
- What is the source of the release? Is the product coming from a pump or piping that will involve a relatively small quantity and risk, or does the release involve a substantial amount of product with the probability of impacting additional exposures (e.g., tank overfill)?
- What is the status of the source of the release? A key task will be source control to isolate the leaking pipe or shut down any pump involved with product movement.
- Is the scenario a spill or fire/spill problem? If the release scenario occurs without ignition, responders should focus on controlling all ignition sources, establishing a foam blanket over the spill area, and providing for air monitoring of the area.
- If ignition has occurred, treat the fire as a large pool fire and cool nearby exposures (storage tank, piping, etc.), as necessary.
- Verify that the dike drains are closed and liquid product is not escaping outside the dike.

## ■ General Methods of Control and Extinguishment

- Ground and dike spills and fires are usually controlled through the use of foam handlines or mobile fire apparatuses. NOTE: Personnel should not physically enter the liquid spill area due to the potential for ignition. (See **FIGURE 6-4**.)

## ■ General Strategy and Tactical Options

The general strategy for managing ground and dike spills and fires is as follows:

1. For fire scenarios, cool critical piping, equipment, and the storage tank as necessary until foam operations are initiated.
2. Extinguish the fire. Small fires may be controlled through the use of dry chemical extinguishers and

**FIGURE 6-4** A fire inside the dike wall is successfully attacked and extinguished with mobile fire apparatus equipped with an articulating foam tower from a safe area.

Refinery Terminal Fire Company.

## CASE STUDY

### FUEL OIL STORAGE TANK FIRE BOILOVER

TACOA, VENEZUELA

DECEMBER 12, 1982

One of the most serious storage tank fires in history occurred on December 19, 1982, in Tacoa, Venezuela, a small town several miles northwest of Caracas. The storage tank supplied fuel to an electric generating plant that supplied power to the city of Caracas. The fuel oil storage tank was approximately 180 feet (54.8 m) in diameter and approximately 56 feet in height. The cone roof tank was built in the late 1970s to store residual fuel oil (#6), and had a weak roof-to-shell seam design for emergency relief venting.

The tank was sited approximately 180 feet (54.8 m) above sea level and was 1,000 feet (304.8 m) from the sea. An adjacent tank of the same design and size, and which stored a similar product, was located 220 feet (67 m) away. Both tanks complied with the requirements of *NFPA 30—Flammable and Combustible Liquids Code* in the three most important areas related to safe tank storage: tank spacing, adequate containment, and adequate emergency relief venting. The tank design, shell-to-shell spacing, and the shell-to-adjacent property spacing met NFPA 30 requirements for non-boilover liquids.

Each tank dike was of sufficient capacity to contain the entire contents of the tank, although the dikes were not intended to contain a boilover. Each tank was equipped with a fixed foam fire protection system consisting of three foam chambers installed in accordance with *NFPA 11—Standard for Low Expansion Foam and Combined Agent Systems*. These tanks also had a pneumatic heat detection system with detectors located just above the vents inside the tanks. Automatic or manual activation of the tanks' heat detection system started the fire pumps and foam system for that tank, and the water cooling system for the other tanks. Three elevated water storage tanks, each with a capacity of 317,000 gallons (1,199,975 L), held the water supply for the fire protection system. Two electric and one diesel pump were available to boost pressure in the water system, when necessary. The water supply was also supplemented by two seawater pumps that could be used to draft from the ocean, if required.

On December 18 at about 11:30 p.m., a high temperature indicator signaled personnel in the plant control room that the oil temperature near the boiler end of a feeding line to the power plant was above normal. The signal was set to activate at 176°F (80°C) and the chart recorder indicated that the temperature was actually 190°F (80°C). Personnel were sent to the tank to shut down one of the two steam heating units operating at that time, and the temperature returned to normal (176°F [80°C]).

On the morning of December 19, at about 6:00 a.m., a three-man crew went to the tank to perform a routine gauging operation. One man remained in a vehicle within the dike, while the other two went to the tank roof. Less than 2 minutes later, a violent explosion occurred which caused the tank roof to blow off and land in the dike, also damaging the water main loop at the base of the tank. The tank burst into flame and a fire ignited within the dike. Although the man in the dike area escaped, his two coworkers died in the explosion.

It was estimated that the product depth in the tank when ignition occurred was 20 feet (6 m). The generating plant had no fire brigade and the nearest fire station was about 20 minutes away. The only access to the site was by means of a winding and narrow road cut into the hillside. All available fire apparatuses was sent to the scene over the next few hours in an attempt to control the fire. Firefighting efforts were hampered by the remoteness of the site, inadequate access, hilly terrain, and damage to the fixed foam fire protection systems. In the initial stages of the incident, individuals in charge at the scene determined that there was little possibility of extinguishing the fire and that the fire would safely burn itself out within the tank and the dike. Therefore, the incident became a "spectator fire" with reporters, fire department personnel, and plant workers standing within 100 to 200 feet (30 to 60 m) of the burning tank, while several firefighters operated a fixed monitor nozzle from the top of the dike, not anticipating the possibility of a boilover.

At approximately 12:15 p.m., a violent boilover blew the tank contents hundreds of feet into the air, forming a giant fireball. Burning oil overflowed the dike. Those individuals who were not killed instantly by the intense radiant heat were caught in a downhill flow of burning oil which reached more than 1,300 feet (396 m) away from the tank, causing some people to jump into the ocean to escape the heat of the fire. Nearby buildings, including about 70 occupied dwellings, ignited and burned. More than 60 vehicles were destroyed, including most of the fire apparatuses on the scene. The boilover claimed the lives of 150 people, including 40 uniformed firefighters and 17 plant employees, and injured scores of civilians. Damage was estimated at $50 million ($124.3 million, in 2018 dollars).

Prior to 1982, there had been no recorded instance of a boilover involving fuel oil #6. Early inquiries into the incident suggested that the product in the tank may not have been fuel oil #6 or that it may have been contaminated. In the past, experimental efforts to ignite such an occurrence have failed. The long-held belief is that fuel oil #6 does not possess the wide range of boiling points necessary to produce the descending hot layer characteristic of a boilover. Furthermore, fuel oil #6 cannot be ignited easily. One of the leading theories concerning the cause of the incident is that it may have been due to tank heating procedures, which raised the temperature of the product higher than necessary and safe, without taking additional precautions. In this scenario, current practices for producing fuel oil #6 may, in fact, result in a product with a boilover ingredient—a wide range of boiling points.

**FIGURE 6-5** Ground fires can often be handled through the use of portable foam handlines.

Courtesy of Gregory G. Noll

quick attack units with the capability to flow both foam and dry chemical agents. Larger fires may be controlled through the use of foam handlines (see **FIGURE 6-5**).

3. Isolate and shut down the source of the release, as appropriate. This may include isolation of valves, diverting product flow, and the shutdown of any transfer pumps. Also ensure that any related tank or dike valves are closed.
4. Conduct air monitoring, as appropriate.

## Full-Surface Tank Fires

Full-surface fires are those which involve the total surface area of the tank, regardless of the type of tank involved (e.g., cone, open floater, covered floater). While the final position of the tank roof (e.g., sunk, partially sunk) will influence firefighting tactics, the full surface of the fuel within the tank is involved in fire. This scenario generally happens in cone roof tanks when the roof is completely lifted off the tank or in floating roof tanks when the floating roof is sunk.

### Size-Up Considerations

Standard operating procedures (SOPs) should be used by the IC to ensure that site safety practices are implemented. If the tank has been preplanned, a copy of the preincident plan should be obtained. Initial questions and tasks that should be addressed as part of the hazard assessment and risk evaluation process include:

- Type of flammable liquid on fire (hydrocarbon or polar solvent).
- Status of tank. Is product being transferred into or out of the tank? Tanks being pumped into will intensify burning at the vent.

### General Methods of Extinguishment

Full-surface tank fires primarily rely upon the use of "over the top" portable foam monitors and nozzles (see **FIGURE 6-6**). The larger the tank diameter (i.e., over 150 ft.), the greater the need for large-volume mobile foam monitors and foam cannons that can apply sufficient foam that will overcome the operational impacts of distance to the tank surface, heat and thermal updraft of the fire, and application onto the fuel surface. Internationally, high-capacity mobile monitor packages have been used to extinguish storage tank fires up to approximately 260 feet (80 meters) in diameter using high-performance fluoro-surfactant–based foam concentrates.

If undamaged and usable, fixed and semifixed foam systems such as foam chambers may also be used to supplement the fire attack process. However, experience shows that in most scenarios involving significant fire or explosion, the foam systems are rendered inoperative.

FIGURE 6-6 Full-surface tank fires primarily rely upon the use of "over the top" portable foam monitors.

Courtesy William T. Hand.

## General Strategy and Tactical Options

As a general guideline, foam application for fire extinguishment should not begin unless there is sufficient foam concentrate onsite to completely control and extinguish the fire. Partial foam applications will be destroyed by the fire. Partial foam application may provide some advantages for rescue situations; however, this action must also be balanced against the increased risk to response personnel. Additional foam will also be required to secure the tank while the product is being pumped out.

The general strategy for managing a full surface tank fire is to:

1. *Protect the structural integrity of the exposed tank shell and any fixed fire protection systems above the fire by using cooling water from master streams or hoselines.*

   The structural integrity of the tank shell can be threatened from inside as well as outside. The lower the product level inside the tank, the greater the possibility that the tank shell will fail due to its exposure to flame impingement. As a general rule, if the tank shell is in direct contact with product inside the tank, there is no immediate risk of structural failure due to heat transfer. Liquid below the burning surface acts as a heat sink and will absorb heat by conduction through the steel tank shell. However, a tank shell exposed to flame impingement above the liquid level will fail rapidly.

   Blistering paint, discoloration of the shell, or steam are indicators that the tank shell requires immediate cooling to prevent sagging or failure. Water should be applied to the exterior of both the involved tank and tank exposures to protect the shell above the burning liquid level. If present and operational, fixed foam connections attached to the tank shell should be protected from fire exposure for the duration of the fire.

   Cooling streams should be applied to adjacent exposure tanks only as needed. Master streams and hoselines should be shut down periodically if steam is not occurring as water contacts with the tank shell. Excess water application will contribute to flooding problems within the tank dike and lower both water pressure and available flows required for firefighting operations.

   Suggested cooling water requirements for exposed tanks include the following:
   - Tanks up to 100 feet in diameter require 500 gpm.
   - Tanks between 100 and 150 feet in diameter require 1,000 gpm.
   - Tanks exceeding the above parameters require 2,000 gpm.

2. *Initiate fire extinguishment operations using offensive and defensive strategies, as appropriate.*

   Foam application rates will vary based upon the fuel and the tank diameter (i.e., surface area). The larger the tank diameter, the greater the required foam application rate. Foam application rates based upon NFPA 11 are as follows:

- Fixed System Hydrocarbon 0.10 gpm/ft$^2$
- Fixed System Seal Protection 0.30 gpm/ft$^2$
- Portable Hydrocarbon Spill 0.10 to 0.16 gpm/ft$^2$
- Portable Hydrocarbon Tank 0.16 gpm/ft$^2$
- Polar Solvents 0.20 gpm/ft$^2$

In addition to NFPA, a number of tank firefighting technical specialists have recommended the following application rates:

- Up to 150 feet diameter 0.16 gpm/ft$^2$
- 151 to 200 feet diameter 0.18 gpm/ft$^2$
- 201 to 250 feet diameter 0.20 gpm/ft$^2$
- 251 to 300 feet diameter 0.22 gpm/ft$^2$
- Over 300 feet diameter 0.24 gpm/ft$^2$

Within the international petroleum industry community, the LASTFIRE Project has reviewed the risks associated with storage tank fires and has developed industry best practices. In a 2005 report, the LASTFIRE Project recommended an application rate of 0.26 gpm/ft$^2$ to its members for large diameter tank fires. Additional information on the LASTFIRE Project can be found at http://www.lastfire.org.uk.

## Cone Roof Tank Fires

### Summary of Construction Features

Cone roof tanks are commonly used for the storage of combustible liquids and some flammable liquids. They are equipped with a cone-shaped or relatively flat roof. If designed to API 650 specifications, the tank will be constructed with a weak roof-to shell seam. In the event of an internal explosion, the roof usually peels back or blows off, leaving the surface of the liquid exposed. Cone roof tanks smaller than 50 feet diameter (sometimes known as dome roof tanks) may not be equipped with a weak roof-to-shell seam. Cone roof tanks also have a pressure/vacuum vent located at the top of the roof to balance the internal and external pressure. (See FIGURE 6-7.)

### General Methods of Extinguishment

There are four primary methods to protect and extinguish a cone roof tank fire. Background information on fixed and semifixed foam protection systems can be found in Chapter 5. These include:

- Foam chambers
- Subsurface and semisubsurface injection
- Portable foam monitors and nozzles
- Portable foam wand

FIGURE 6-7 Cone roof tanks are equipped with a cone-shaped or relatively flat roof.

Courtesy of William T. Hand.

#### Problem: Vent Fires

Vent fires are among the most common types of cone roof storage tank emergencies. They are usually caused by lightning strikes or by the ignition of vapors in the vicinity of the vent due to improper welding or cutting operations. When a vent catches fire, a large flame can burn up and away from the roof or it can "lay down" onto the roof due to high winds and create an exposure problem. (See FIGURE 6-8.)

### Size-Up Considerations

Standard operating procedures (SOPs) should be used by the IC to ensure that site safety practices are implemented. If the tank has been preplanned, a copy of the preincident plan should be obtained. Initial questions and tasks that should be addressed as part of the hazard assessment and risk evaluation process include:

- Type of flammable liquid on fire (hydrocarbon or polar solvent).
- Vapor air mixture in the tank and the potential for flashback through the tank pressure-vacuum vent. A fire at a tank vent with a snapping blue-red, nearly smokeless flame indicates that the vapor/air mixture in the tank is flammable (explosive). As long as the tank is breathing out through the pressure-vacuum vent, the flame cannot flash back

FIGURE 6-8 Vent fire on a cone roof tank.

Courtesy of Gregory G. Noll.

into the tank because of the positive pressure, high velocity flow through the vent.

- Status of tank. Is product being transferred into or out of the tank? Tanks being pumped into will intensify burning at the vent.
- Flame impingement from the vent fire onto the tank shell or tank roof.
- Ability to safely access the roof by ladders and alternative routes of escape from the roof.
- Slope of the tank roof.
- Structural integrity of the tank roof. Is the roof safe to walk on (e.g., evidence of corrosion)? As a general rule, responders should avoid walking on cone roofs.

## General Strategy and Tactical Options

The general strategy for managing a tank vent fire is to:

1. Lower the internal tank pressure by cooling the external tank shell with water so that the pressure-vacuum (PV) vent resets and extinguishes the vent fire. If the tank shell is exposed to fire, cooling water should be applied to the roof and shell. A pressure reduction in the tank caused by cooling may extinguish the fire when the PV vent closes.
2. Extinguish using hose streams. Vent fires have been successfully extinguished by applying a hose stream to the fire and discharging a dry chemical (e.g., Purple K) into the stream. The dry chemical is carried to the fire by the water stream, and together they extinguish the fire. After extinguishment has been achieved, air monitoring must be performed to ensure that there are no flammability or toxicity hazards present. Air monitoring should continue until the emergency is terminated by the IC. Additional monitoring may be required as the site is being restored to normal operation.
3. Introduce an inert gas (e.g., nitrogen) into the tank to create an oxygen-deficient atmosphere in the vapor space above the liquid level; OR
4. Close the pressure/vacuum snuffer plate (if equipped); OR
5. Use aerial apparatuses to access and extinguish the fire using handheld dry chemical extinguishers (e.g., Purple K). Any attempt to extinguish a vent fire using dry chemical must take into consideration that the PV vent may have been damaged by heat from the fire and will not close after the fire has been extinguished. A PV vent in the *failed open position* could enable flashback into the tank if the fire is initially extinguished with dry chemical and reignites. Flashback into a tank containing a flammable vapor/air mixture could produce an internal tank explosion. Remember, tank vent fires burning with a yellow-orange flame and emitting black smoke indicate that the vapor/air mixture in the tank is above the flammable (explosive) limit. A snapping blue-red, nearly smokeless flame indicates that the vapor/air mixture in the tank is flammable (explosive).

### Problem: Roof Partially or Completely Separated

If the vapor space in the cone roof tank is in the explosive range at the time of ignition, the subsequent explosion should separate the roof from the tank shell at the weak roof-to-shell seam (API 650 design) and likely result in a full-surface fire. The roof may separate in one piece or in fragments, or it may peel back over the side of the tank and still be attached to the tank shell in one or more places. Fragments can travel great distances and cause additional damage to other tanks, pipe racks, etc. The possibility exists that other flammable or combustible leaks may occur in the vicinity of the tank fire.

The roof may also lift into the air and fall back into the tank, or peel back and hang over the side of the tank. If the roof is partially attached to the tank shell, it may restrict foam application. When this occurs the surface of the burning liquid is obstructed by the wreckage, making fire extinguishment difficult. If foam chambers are in place, they may also be damaged by the initial explosion or by the roof. (See FIGURE 6-9.)

## Size-Up Considerations

Standard operating procedures (SOPs) should be used by the IC to ensure that site safety practices are implemented. If the tank has been preplanned, a copy of the preincident plan should be obtained. Initial questions and tasks that

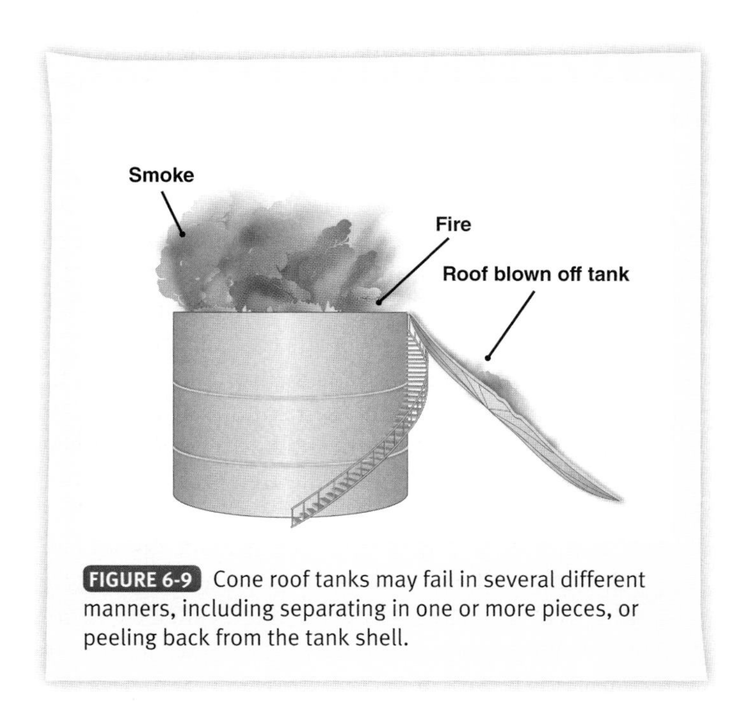


FIGURE 6-9 Cone roof tanks may fail in several different manners, including separating in one or more pieces, or peeling back from the tank shell.

should be addressed as part of the hazard assessment and risk evaluation process include:

- Type of flammable liquid on fire (hydrocarbon or polar solvent).
- Flash point of the flammable liquid (high or low flash point).
- Level of product in the tank. For example, is the upper shell exposed to fire? If so, the shell may curl back into the tank and further complicate fire extinguishment.
- Status of tank. For example, was product being transferred into or out of the tank at the time of the explosion? Are all pumping operations shut down? As a general rule of thumb, pumping operations should be terminated as soon as possible until an initial tactical action plan is developed.
- Probability that the fire will be confined to its present size. Has the roof damaged the tank shell of the tank on fire or an adjacent tank?
- Three-dimensional fires inside the diked area from flanges or pumps.
- Flame impingement on the external shell, piping, valves, and flanges from a dike fire.
- Availability of sufficient foam to extinguish the fire.
- Position of the separated tank roof (e.g., is the roof still attached to the shell?).
- Status of dike drains or pipeline penetrations through the dike wall.
- Exposures to adjacent tankage (e.g., LPG storage).
- Weather conditions (e.g., high winds, lightning, wind direction, freezing conditions).

## General Strategy and Tactical Options

The general strategy for managing a cone roof tank fire with the roof partially separated and the surface of the liquid totally involved is to:

1. *Protect the structural integrity of the exposed tank shell and any fixed fire protection systems above the fire by using cooling water from master streams or hoselines.*

   The structural integrity of the tank shell can be threatened from inside as well as outside. The lower the product level inside the tank, the greater the possibility that the tank shell will fail due to its exposure to flame impingement on the inside. As a general rule, if the tank shell is in direct contact with product inside the tank, there is no immediate risk of structural failure due to heat transfer. Liquid below the burning surface acts as a heat sink and will absorb heat by conduction through the steel tank shell. However, a tank shell exposed to flame impingement above the liquid level will fail rapidly.

   Blistering paint, discoloration of the shell, or steam are indicators that the tank shell requires immediate cooling to prevent sagging or failure. Water should be applied to the exterior of the tank to protect the shell above the burning liquid level. Fixed foam connections attached to the tank shell should be protected from fire exposure for the duration of the fire. Master streams and hoselines should be shut down periodically if steam is not occurring as water contacts with the tank shell. Excess water application will contribute to flooding problems and lower water pressure required for firefighting operations.
2. *Extinguish the tank fire using fixed systems or master streams and hoselines.*

   **Subsurface Application.** Subsurface injection of foam into hydrocarbon storage tanks requires that foam be injected at least 1 foot above any water bottom in the tank. Additional water bottom will be created during subsurface injection and may require that water be drawn off the bottom of the tank before and during injection.

   Special high-backpressure foam makers will be required for subsurface injection. Depending on its design, the foam maker will normally tolerate 20% to 40% of its inlet pressure in backpressure against the discharge of the foam maker. The pressure generally consists of the head pressure in the tank and the friction loss of the finished foam in the piping between the foam maker and the injection point. Special hydraulic calculations will be required prior to initiating firefighting.

   The flow velocity of the foam in the inlet piping to the tank is critical. For flammable liquids, the velocity should not exceed 10 feet per second, based on the expanded volumetric rate. For combustible liquids, a velocity of up to 20 feet per second should not be exceeded. See *API-2021, Appendix E* for graphs which indicate friction loss characteristics and inlet velocities for various pipe sizes.

   Subsurface injection may require multiple injection points depending on the product involved and the diameter of the tank. Subsurface injection usually extinguishes the center of a tank fire first and the rim area last. Elevated platforms may provide a good vantage point to evaluate progress of fire extinguishment.

   **Topside Application.** Topside application of firefighting foams may be accomplished using several methods of application, including fixed foam systems, foam wands, ground level high capacity foam monitor nozzles, and mobile aerial devices equipped with foam nozzles. (See **FIGURE 6-10**.) See the Full-Surface Tank Fires section of this chapter for detailed information on topside application rates and operational requirements.

FIGURE 6-10 Topside application can be accomplished using fixed foam systems, ground monitors, and mobile aerial devices. This is an example of a fixed foam suppression foam chamber.

Courtesy of William T. Hand.

FIGURE 6-11 Open top floating roof tanks are easily identified by the wind girder at the top of the tank and the roof stairway.

Courtesy of Gregory G. Noll.

## Open Top Floating Roof Tank Fires

### Summary of Construction Features

Open top floating roof tanks are similar to the cone roof tank, except that the roof is not fixed to the tank shell and actually floats on top of the liquid inside the tank. This is made possible by pontoons or flotation built into the roof. Floating on the liquid surface, the roof can move up and down inside the tank as the liquid level changes. The only time an open floating roof tank will have a vapor space is when the liquid level is in a low position below the floating roof supports attached to the tank floor. Seals are in place between the tank shell and the floating roof. (See FIGURE 6-11.)

### General Methods of Extinguishment

There are four primary methods to protect and extinguish an open floating roof tank fire. Background information on fixed and semifixed foam protection systems can be found in Chapter 5. These include:

- Foam chambers
- Catenary and coflexip systems, which discharge foam directly from the floating roof into the seal area
- Portable foam monitors
- Portable foam handlines applied from elevated locations

#### Tactical Problem: Rim Seal Fire

Seal fires are the most common type of emergency on open top floating roof tanks. Seal fires are usually caused by a lightning strike near the top of the tank. When the seal area catches fire, both flammable vapors and the seal itself burn. Eventually, the fire will continue to burn into the seal until the entire seal area ringing the floating roof is on fire. (See FIGURE 6-12.)

Seal fires may burn for some time with little damage to the tank if the upper shell is kept cool from outside the tank above the liquid level. In contrast, seal fires can also damage the roof beyond repair, taking the tank out of service. Seal fires can damage pontoons, causing the pontoons to fail and the roof to partially sink and expose a larger fuel surface to the fire. Seal fires may also damage and slowly burn into aluminum honeycombed metal deck roofs.

Seal fires normally don't become full-surface fires without some help! Responders should ensure that firefighting operations from the ground do not accidentally sink the floating roof.

### Size-Up Considerations

Standard operating procedures (SOPs) should be used by the IC to ensure that site safety practices are implemented. If the tank has been preplanned, a copy of the preincident plan should be obtained. Initial questions and tasks that should be addressed as part of the hazard assessment and risk evaluation process include:

- Type of flammable liquid on fire (hydrocarbon or polar solvent).
- Level and position of the tank roof. For example, is the floating roof in a high or low position? Is the roof resting on its support legs?
- Status of tank. Is product being transferred into or out of the tank?
- Probability that the fire will be confined to its present size.

FIGURE 6-12 Open top floating roof tank seal fires usually begin in one area and then spread across the tank throughout the entire seal area. From ground level it may appear that the entire roof is on fire. An elevated aerial ladder or platform will identify the scope of the problem.

Courtesy of John Hess.

- Flame impingement on the internal tank shell above the floating roof level.
- Status of the pontoon roof. For example, is the roof flooded with water? Are the pontoons still intact or have they been compromised?
- Safe access to the roof to size up the extent of the seal fire. Are there alternative routes of escape from the roof? Sending firefighting crews onto a floating roof with handlines for extinguishment is undesirable. Seal fire conditions may change rapidly and cut off the only route of escape. A floating roof lower than 8 feet from the top of the tank shell must be treated as a confined space.
- Type of seals in place (e.g., fabric type, primary and secondary).
- Structural integrity of the wind girder. Can the wind girder be used as a safe walking surface?
- Weather conditions, such as high winds, wind direction, lightning, etc.

## General Strategy and Tactical Options

The general strategy for managing rim seal fires on open floating roof tanks is to:

1. Protect the structural integrity of the exposed tank shell above the seal fire by using cooling water from master streams or hoselines. Water should be applied to the outside tank shell using care not to apply water over the top of the tank onto the floating roof. Flooding the roof can result in sinking the roof and can thus cause a seal fire to quickly become a full-surface fire.
2. Extinguish the seal fire using fixed fire protection systems (e.g., foam chambers, catenary system, etc.); OR
3. Extinguish the seal fire with portable and/or mobile fire equipment from a safe elevated position.

Open top floating roof tank seal fires are generally confined to the annular seal area between the floating roof and the tank shell. The following methods may be used to extinguish the fire.

### Foam Chambers

This method requires that a permanently fixed foam dam be in place between the seal area and the tank shell. If a foam dam is not in place or there is a hole in the dam, foam will flow out of the seal area and onto the roof, possibly sinking the floating roof. Foam dams are normally 12 to 24 inches in height and must extend above the seal area. Foam application rates average 0.30 to 0.50 gpm/ft$^2$ depending on the type of seal. Foam and tank manufacturer guidelines should be consulted and integrated into the tank preincident plan. At least 20 minutes of foam supply should be on hand before beginning the fire attack. (See FIGURE 6-13.)

### Catenary System

Catenary systems are designed to apply foam at evenly spaced points on the roof above or below the seal. A foam

FIGURE 6-13 Foam chambers allow foam to flow from a foam maker at ground level through vertical piping systems and then flow down the inside tank wall into the foam seal. A foam dam prevents the foam from flowing onto the floating roof.

Courtesy of Gregory G. Noll.

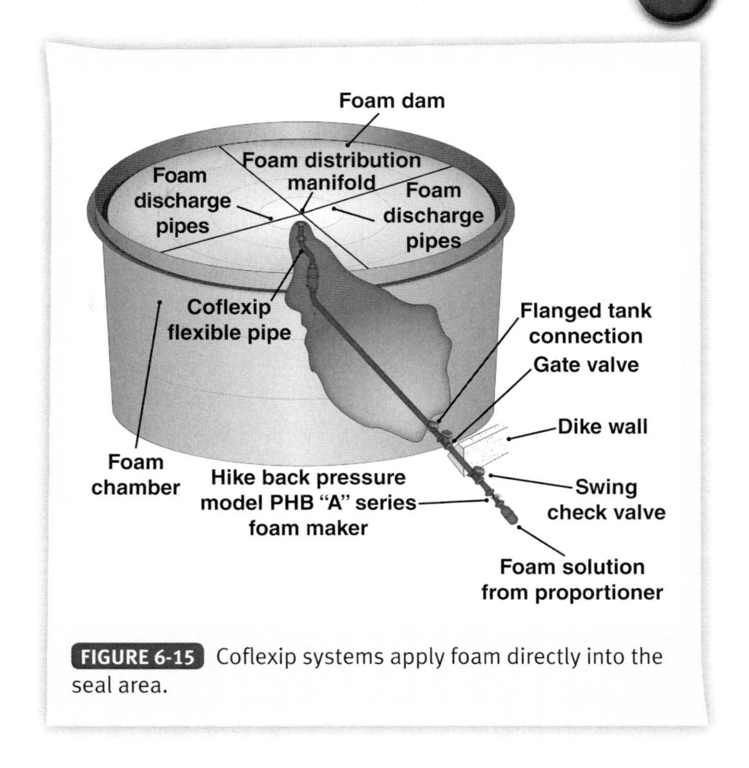


FIGURE 6-15 Coflexip systems apply foam directly into the seal area.

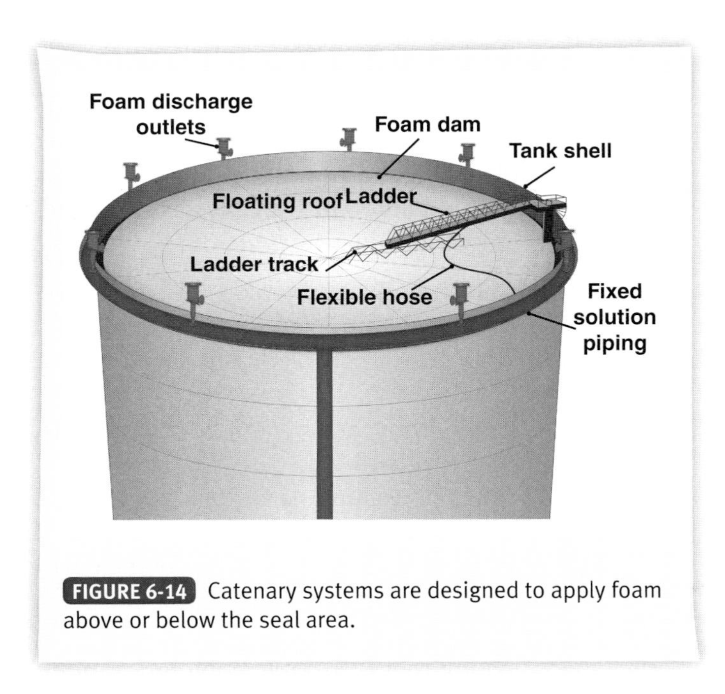


FIGURE 6-14 Catenary systems are designed to apply foam above or below the seal area.

dam may be required depending on the type of seal. Foam application rates average 0.30 to 0.50 gpm/ft$^2$ depending on the type of seal. Foam and tank manufacturer guidelines should be consulted and integrated into the tank preincident plan. At least 20 minutes of foam supply should be on hand before beginning the fire attack. (See FIGURE 6-14.)

### Coflexip System

This system uses either a high-backpressure foam maker located outside the tank or a foam maker mounted on the floating roof. The foam solution enters the bottom of the tank through fixed piping. The pipe carries the foam solution to a distribution manifold mounted at the center of the floating roof. This manifold is equipped with a blowout plug to prevent vapors from entering the foam piping. During foaming operations, the plug is forced out and the foam solution passes from the manifold through fixed discharge outlets located at the seal area on the perimeter of the tank. The flexible pipe inside the tank adjusts to the rise and fall of the tank.

Foam application rates average 0.30 to 0.50 gpm/ft$^2$ depending on the type of seal. Foam and tank manufacturer guidelines should be consulted and integrated into the tank preincident plan. At least 20 minutes of foam supply should be on hand before beginning the fire attack. (See FIGURE 6-15.)

### Foam Standpipe Handlines

This system consists of a riser that extends up the outside of the tank shell to the area near the top of the stairway. A single, fixed chamber is required at the top of the stairs so that foam can be discharged onto the seal area directly below the platform and provide responders with a safe base of operations. Firefighters should always take a second charged hoseline while ascending the stairway for personal protection. The fixed standpipe connection on the stairway is then used to supply up to two foam handlines. Firefighters must then carry the foam handline down the stairway or around the wind girder to discharge foam into the seal area.

### Portable Foam Handlines

If a fixed protection system is not in place, it will be necessary to fight the fire with portable and/or mobile equipment. Foam streams may be directed inside the tank onto the shell above the seal area from the gauger platform, an adjacent tank or catwalk, or an elevated aerial platform or ladder. Flowing large straight streams of foam or water directly into the damaged seal area should be avoided, as it may splash product onto the roof and increase the fire's intensity. Foam should be applied to the inside of the tank shell so that it can run down into the seal area.

Placing firefighting crews onto a floating roof with handlines for extinguishment is undesirable. Seal fire conditions may change rapidly and cut off the only route of escape. A floating roof lower than 8 feet from the top must be treated as a confined space.

Firefighting operations conducted from the gauger platform or the wind girder must be conducted with special care. Wind, slippery walking surfaces, and cumbersome protective clothing and equipment make firefighting from elevated positions extremely hazardous. Hand rails or rigged fall protection should be in place to protect personnel. Firefighters should have more than one way off any elevated work area (e.g., tank stairs or an elevated aerial platform or ladder). Small seal fires can grow in intensity and block escape routes. Firefighters in elevated positions must have two-way radio communications with ground units. Firefighting operations should be continuously observed from a safe and secure elevated position. If the seal fire has been burning for some time, it may be necessary to apply cooling water to the outside of the tank shell. The need for cooling water can be determined by blistering or discolored paint. Cooling water streams should be shut down when steaming stops.

### Tactical Problem: Partially Sunken Roof

Floating roofs on both open and covered floating roof tanks can partially sink. This may be caused by failure of the pontoons providing flotation for the roof or from an excess load created by rain or snow. The roof may also be sunk by pouring too much water onto the roof during firefighting operations (not good for career advancement).

While some floating roofs have been known to completely sink below the liquid surface of the tank, experience shows that they often sink partially. In this configuration, one side of the floating roof is cocked against the tank wall in an elevated position above the liquid level, while the other half of the floating roof is submerged.

The primary problem created by a partially sunken roof is the void space created below the roof between the liquid level and the underside of the floating roof. Foam applied to the surface of the burning liquid will be obstructed by the roof and cannot flow onto the liquid under the void space. (See FIGURE 6-16.)

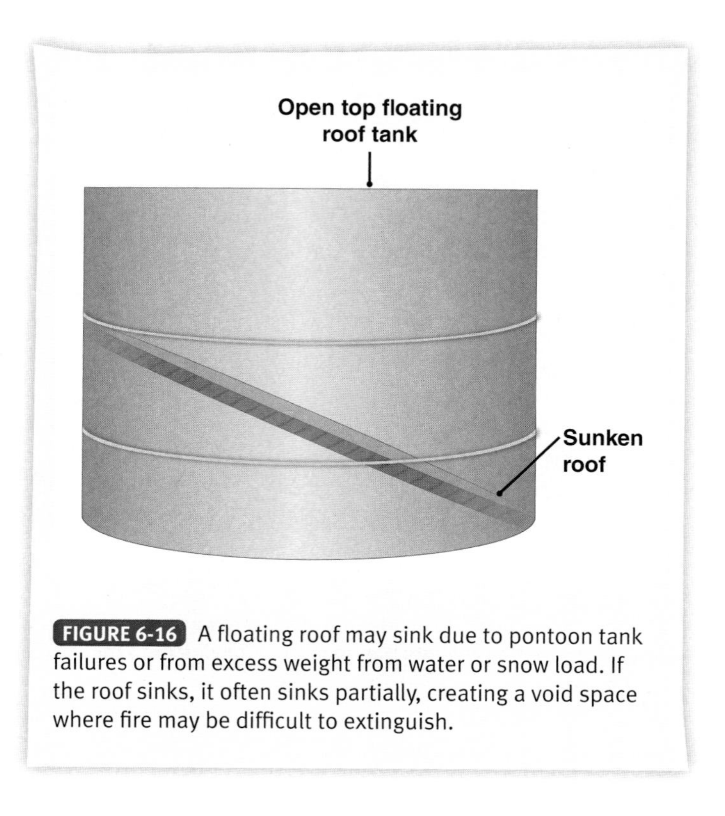


**FIGURE 6-16** A floating roof may sink due to pontoon tank failures or from excess weight from water or snow load. If the roof sinks, it often sinks partially, creating a void space where fire may be difficult to extinguish.

## Size-Up Considerations

Standard operating procedures (SOPs) should be used by the IC to ensure that site safety practices are implemented. If the tank has been preplanned, a copy of the preincident plan should be obtained. Initial questions and tasks that should be addressed as part of the hazard assessment and risk evaluation process include:

- Type of flammable liquid involved (e.g., sour crude oil).
- Flash point of the flammable liquid (high or low flash point).
- Product level within the tank. How much of the upper shell is exposed?
- Status of tank. For example, is product being transferred into or out of the tank?
- Damage to the tank floor from the legs of the floating roof.
- Is the roof partially or completely sunk?
- Type of floating roof (e.g., pan, pontoon, or honeycomb roof).
- Type, nature, and degree of roof damage, and the status of roof piping.
- Availability of sufficient foam to provide vapor suppression.
- Status of dike drains or penetrations through the dike wall.
- Weather conditions, such as high winds, wind direction, lightning, freezing conditions, etc.
- Possibility that other adjacent roofs will sink due to heavy rain/snow pack.

## General Strategy and Tactical Options

The general strategy for managing a fully or partially sunken floating roof is to:

1. Stop product flow into the tank and verify shutdown.
2. Apply foam to the fuel surface using fixed systems or portable or mobile fire apparatus.
3. Check for tank floor leaks caused by the cocked floating roof puncturing the bottom. Where present, control flammable liquid spills inside the diked area and apply foam as required.
4. Maintain vapor suppression using a foam blanket until the tank inventory has been pumped out. If available, a foam concentrate with a long drainage time should be used.

The primary tactical objective should be to maintain vapor suppression, prevent ignition of the exposed flammable liquid, and maintain the integrity of the foam blanket for an extended period as the tank's contents are pumped out. The method of foam application will depend on availability of application devices (e.g., fixed systems, portable devices, mobile apparatus). Although any application method is acceptable, rapid foam application is critical to prevent ignition.

# Covered Floating Roof Tank Fires

## Summary of Construction Features

A covered floating roof tank is a combination of a cone roof tank and an open floating roof tank. It has a fixed cone roof and an internal pan or deck-type floating roof that rides directly on the product surface. A version of a covered floating roof tank uses a geodesic dome rather than the cone roof. Geodesic dome roofs are often retrofitted onto open top floating roof tanks to convert them to covered floating roof tanks. Seals are in place between the tank shell and the floating roof. Open vents are provided around the tank shell between the fixed roof and the floating roof. The external fixed roof has a topside manway for access to the floating roof, although this is of little value during an emergency. (See FIGURE 6-17.)

## General Methods of Extinguishment

Methods for extinguishing an internal floating roof tank fire will primarily depend upon the type of fire. Potential fire scenarios can include seal fires, surface fires with the internal floating roof partially sunk, and a full-surface fire where the fixed roof has been blown off and the internal pan roof sunk.

Background information on fixed and semifixed foam protection systems can be found in Chapter 5. Extinguishing options include the following:

FIGURE 6-17 Covered floating roof tanks can easily be distinguished from an open top floating roof tank by identifying the "eyebrow vents" at the top of the tank.

Courtesy of Jorge Carrasco.

Seal Fires (Roof Still in Place and Undamaged)

- Fixed foam suppression systems with through-the-roof piping connected to foam nozzles or poring devices that flow foam down the tank wall and into the seal area.

Full Surface Fires (Roof Damaged or Removed)

- Foam chambers
- Portable foam monitors
- Foam wand
- Aerial devices

Subsurface injection is not practical on a covered floating roof tank because the internal floating roof would not allow the foam to reach the surface of the fire.

### Tactical Problem: Seal Fire Below the Roof

Internal floating roof tanks have an excellent fire safety record. The vapor space between the external and internal roof is usually vapor free from ignitable flammable vapor/air mixtures, except during periods of initial fill and for 18 to 25 hours thereafter, depending on the volatility of the product. (See FIGURE 6-18.)

The primary tactical problem with an internal floating roof tank fire is that the fixed roof prevents direct access to the internal floating roof seal area. While seal fires in these tanks are rare, they can be very difficult to extinguish if fixed foam systems are not installed or are working improperly.

FIGURE 6-18 Covered floating roof tank fires present challenges for extinguishment operations using portable foam monitors.

Courtesy of Gregory G. Noll.

## Size-Up Considerations

Standard operating procedures (SOPs) should be used by the IC to ensure that site safety practices are implemented. If the tank has been preplanned, a copy of the preincident plan should be obtained. Initial questions and tasks that should be addressed as part of the hazard assessment and risk evaluation process include:

- Type of flammable liquid involved (e.g., sour crude oil).
- Level and position of the floating roof. For example, is the floating roof resting on its support legs?
- Floating roof design (e.g., aluminum, fiberglass pan).
- Status of tank. For example, is product being transferred into or out of the tank? All product flow to and from the tank should be stopped until the situation can be completely assessed.
- Probability that the fire will be confined to its present size.
- Flame impingement on the internal tank shell above the roof level.
- Availability of sufficient foam to extinguish the seal fire.
- Weak roof-to-shell seam (e.g., API 650 construction).
- Type of primary and secondary seals installed.

## General Strategy and Tactical Options

The general strategy for managing an internal floating roof tank rim seal fire is to:

1. Protect the structural integrity of the exposed tank shell above the seal fire by using cooling water from master streams or hoselines.
2. Extinguish any dike fires.
3. Extinguish the seal fire using fixed fire protection systems, such as foam chambers or rimseal pouring nozzles installed below the roof but above the maximum level of the floating roof. Foam flows down the inside of the tank wall and into the seal area.

Internal floating roof tank fires are extremely difficult to extinguish unless the tank is equipped with a fixed foam system. Seal fires or rim fires are difficult to fight with portable equipment. The side vents are too small to permit access by foam streams directed from ground level. Elevated aerial devices may be able to access vents; however, heavy smoke conditions make directing streams difficult. In addition, tank location and other obstructions (e.g., power lines) may make apparatus placement difficult. The possibility exists that misdirected foam streams may sink the internal floating roof.

These fires are generally confined to the annular seal area between the floating roof and the tank shell. The following methods may be used to extinguish the fire:

1. Fixed foam systems (e.g., foam chamber) offer the greatest opportunity for successful fire extinguishment.
2. Specially designed foam towers and foam wands have been used successfully to apply foam through the tank vent.
3. If fixed fire protection is not in place, it will be necessary to adopt a defensive or nonintervention tactical option.

Firefighting should be continuously observed from a safe and secure elevated position. If a strategy is adopted and it appears that it is not being successful, it should be shut down and alternative options evaluated.

If the seal fire has been burning for some time, it may be necessary to apply cooling water to the outside of the tank shell. The need for cooling water can be determined by blistering or discolored paint. Cooling water streams should be shut down when steaming stops.

# Horizontal and Vertical Low Pressure Tank Fires

## Summary of Construction Features

Horizontal and vertical low pressure storage tanks are usually designed to the minimum requirements of *API 620—Carbon Steel Aboveground Storage Tanks*. Tanks constructed to API 620 standards handle internal pressures not exceeding 15 psig and are intended to hold or store liquids with vapors above the surface of the liquid or to store vapors without liquid in the tank. (See FIGURE 6-19.)

Tanks meeting these specifications are cylindrically shaped with a top and a bottom for the vertical tank, or two ends for the horizontal tank. The horizontal tank is

**FIGURE 6-19** When a vent fire occurs on a low pressure storage tank, the upper tank shell needs to be kept cool to prevent escalation of the fire and potential tank failure.

Courtesy of Gregory G. Noll.

**FIGURE 6-20** Horizontal and vertical low pressure storage tanks can be found almost anywhere.

Courtesy of Gregory G. Noll.

usually mounted on concrete supports, although there are many older versions still in service which are mounted on structural steel. Both versions are equipped with a pressure/vacuum vent.

### Tactical Problem: Vent Fire Without a Spill

Most vent fires on horizontal and vertical low pressure tanks are caused by lightning strikes. The fire may burn lazily above the vent, or with intensity depending on the temperature of the liquid, its vapor pressure, and the intensity and direction of the wind. The basic tactical problem with a vent fire is to keep the upper half of the exposed tank shell cool until the vent fire can be extinguished. There are two potential dynamics occurring: (1) Flame impingement on the tank shell will weaken the steel, and (2) the internal tank pressure will increase and intensify the vent fire and may eventually result in tank rupture. (See **FIGURE 6-20**.)

## Size-Up Considerations

Standard operating procedures (SOPs) should be used by the IC to ensure that site safety practices are implemented. If the tank has been preplanned, a copy of the preincident plan should be obtained. Initial questions and tasks that should be addressed as part of the hazard assessment and risk evaluation process include:

- Is the tank of welded steel or bolted construction?
- Is the tank elevated, on the ground, or partially covered by soil?
- Is the tank horizontal or vertical?
- If the tank is horizontal, what type of structural supports are in place? Are they protected against spill fires? Unprotected structural steel supports are only as strong as their weakest point (i.e., connectors) and can fail early in the fire, causing the tank to breach.
- Where is the fire impinging on the tank?

## General Strategy and Tactical Options

The general strategy for managing a vent fire on a vertical or horizontal low pressure storage tank is to:

1. Protect the structural integrity of the exposed tank shell by using cooling water from master streams or hoselines.
2. Extinguish any spill fires.
3. Extinguish the vent fire using a firehose stream or a handheld portable fire extinguisher.

### Tactical Problem: Tank Overfill

Many low pressure storage tank fires have been caused by overfilling the tank. The product overflow may come through the tank vent or, in the case of a horizontal tank at a terminal, through an open product fill line.

## Size-Up Considerations

Many firefighters have been killed or injured while attempting to attack and extinguish low pressure horizontal or vertical storage tank fires. In a spill situation, the fire can quickly impinge upon the vapor space of the tank and cause it to fail violently. Remember that a PV vent is designed for normal transfer operations and does not provide adequate emergency pressure relief.

As with pressure relief devices on LPG tank containers, the fact that the relief valve is functioning and relieving internal tank pressure does not necessarily mean the tank won't fail.

Standard operating procedures (SOPs) should be used by the IC to ensure that site safety practices are implemented. If the tank has been preplanned, a copy of the preincident plan should be obtained. Initial questions and tasks that should be addressed as part of the hazard assessment and risk evaluation process include:

- Is the tank of welded steel or bolted construction?
- Is the tank horizontal or vertical?
- If the tank is horizontal, what type of structural supports are in place? Are they protected against spill fires? Unprotected structural steel supports are only as strong as their weakest point (i.e., connectors) and can fail early in the fire, causing the tank to breach.
- Where is the spill fire impinging on the tank?
- Is the PV valve functioning? If so, is the fire intense? This would indicate high pressure inside the tank and is an early warning sign of potential violent tank failure. However, it is not a reliable way to determine *when* to withdraw from the area.
- Is the spill flowing away from the burning tank and creating other exposure problems?
- Is the tank diked? Low pressure tanks in older facilities and in some rural areas have been found with inadequately constructed dikes (that means they leak!).
- Are the dike drains open?

## General Strategy and Tactical Options

The general strategy for managing a spill fire involving a vertical or horizontal low pressure storage tank is to:

1. Protect the structural integrity of the exposed tank shell and the structural supports by using cooling water from master streams or hoselines from a safe distance.
2. Extinguish the spill fire using master streams or handlines.
3. Extinguish the vent fire (if any) using a firehose stream, dry chemical (Purple K), or twin agent units (e.g., Purple K and AFFF).

## Summary

While aboveground storage tank emergencies are low probability events, they often have high consequences. Events do occur annually in the United States and throughout the world. The two most common causes of aboveground storage tank fires are lightning strikes and tank overfills due to human error.

Storage tank emergencies can present the incident commander with multiple tactical problems that must be sorted out and prioritized. The primary hazards and risks associated with storage tank emergencies include confined spaces, hydrogen sulfide, and boilover.

Dikes are designed to contain flammable and combustible liquids in the event of an accident, such as an overfill. The larger the tank capacity, the larger the diked area. As a general safe operating practice, diked areas should be treated as confined spaces during firefighting operations.

Cone roof tanks are commonly used for the storage of combustible liquids and some flammable liquids. They are equipped with a cone-shaped or relatively flat roof. If designed to API 650 specifications, the tank will be constructed with a weak roof-to shell seam. Vent fires are among the most common types of cone roof storage tank emergencies. They are usually caused by lightning strikes or by the ignition of vapors in the vicinity of the vent due to improper welding or cutting operations.

Open top floating roof tanks are similar to the cone roof tank, except that the roof is not fixed to the tank shell and actually floats on top of the liquid inside the tank. This is made possible by pontoons or flotation built into the roof. Floating on the liquid surface, the roof can move up and down inside the tank as the liquid level changes. Seal fires are the most common type of emergency on open top floating roof tanks. Seal fires are usually caused by a lightning strike near the top of the tank. Floating roof tanks can partially sink due to failure of the pontoons providing flotation for the roof or from an excess load created by rain or snow. The primary problem created by a partially sunken roof is the void space created below the roof between the liquid level and the underside of the floating roof.

Covered floating roof tanks are a combination of a cone roof tank and an open floating roof tank. It has a fixed cone roof and an internal pan or deck-type floating roof that rides directly on the product surface. Another version is a geodesic dome roof rather than the cone roof. Seals are in place between the tank shell and the floating roof. Open vents are provided around the tank shell between the fixed roof and the floating roof. Like the open top floating roof tank, seal fires are the most common type of fire involving the covered floating roof tank. While the risk of a seal fire is low, such fires can be difficult to extinguish unless a fixed fire suppression is in place.

Horizontal and vertical low pressure storage tanks are designed to handle internal pressures not exceeding 15 psig and are intended to hold or store liquids with vapors above the surface of the liquid or to store vapors without liquid in the tank. Tanks meeting these specifications are cylindrically shaped with a top and a bottom for the vertical tank, or two ends for the horizontal tank. The horizontal tank is usually mounted on concrete supports, although there are many older versions still in service which are mounted on structural steel. Both versions are equipped with a pressure/vacuum vent. Vent fires and tank overfills are the most common emergencies on these tanks.

## References and Suggested Readings


1. American Petroleum Institute, *Guidelines for Work in Inert Confined Spaces in the Petroleum Industry and Petrochemical Facilities* (4th edition), API Publication 2217A, Washington, DC: API (July 2009).
2. American Petroleum Institute, *Management of Atmospheric Storage Tank Fires* (4th edition), API Publication 2021, Washington, DC: API (Reaffirmed May 2006).
3. American Petroleum Institute, *Prevention and Suppression of Fires in Large Aboveground Atmospheric Storage Tanks* (1st edition), API Publication 2021A, Washington, DC: API (1998).
4. American Petroleum Institute, *Safe Access/Egress Involving Floating Roofs of Storage Tanks in Petroleum Service* (2nd edition), API Publication 2026, Washington, DC: API (Reaffirmed June 2006).
5. Campbell, Richard, *Fires at Outside Storage Tanks*, Quincy, MA: National Fire Protection Association (2008).
6. Chang, James, and Cheng-Chung Lin, A Study of Storage Tank Accidents, *Journal of Loss Prevention in the Process Industries* (2006), pp. 51–59.
7. French Ministry of Environment, *Boilover of a Crude Oil Tank, 30 August 1983 at Milford Haven, Wales, United Kingdom* (2008).
8. LASTFIRE Boilover Research Position Paper and Practical Lessons Learned, Aylesbury, Bucks, UK: LASTFIRE (December 3, 2016).
9. Noll, Gregory G., and Michael S. Hildebrand, *Hazardous Materials: Managing the Incident* (4th edition), Burlington, MA: Jones and Bartlett Learning (2014), pp. 352–365.
10. Persson, Henry, and Anders Lonnermark, *Tank Fires: Review of Fire Incidents 1951–2003*, Brandforsk Project 513-021, SP Swedish National Testing and Research Institute (2004).
11. Shelley, C.H., Petroleum Storage Tank Facilities–Part 2, *International Firefighter* (April 2015).
12. Shelley, C.H., *Storage Tank Fires: Is Your Department Ready?* Tulsa, OK: Fire Engineering (2007).
13. Shelley, C.H., A.R. Cole, and T.E. Markley, *Industrial Firefighting for Municipal Firefighters,* Tulsa, OK: Fire Engineering (2007).
14. Williams, Dwight P., Over the Top: Techniques and Logistics for Extinguishing Large Tank Fires, *Industrial Fire Safety* (November/December, 1992), pp. 21–25.

# Glossary

Aboveground Bulk Storage Tank A horizontal or vertical tank that is listed and intended for fixed installation, without backfill, above or below grade, and is used within the scope of its approval or listing. (NFPA 30.A)

Alcohol-Resistant Concentrate (ARC) This and aqueous film-forming foams (AFFFs) are Class B firefighting foams that can be applied to both hydrocarbons and polar solvents. They can be applied at various concentrations, including 3% hydrocarbon to 3% polar solvent (known as 3 × 3 concentrates), and 3% hydrocarbon to 6% polar solvent (known as 3 × 6 concentrates). When applied to a polar solvent fuel, they will often create a polymeric membrane rather than a film over the fuel. This membrane separates the water in the foam blanket from the attack of the polar solvent. Then, the blanket acts in much the same manner as a regular AFFF.

American Chemistry Council The parent organization that operates CHEMTREC™.

American National Standards Institute (ANSI) A 501(c)3 private, not-for-profit organization that oversees the creation, promulgation, and use of thousands of norms and guidelines that directly impact businesses in nearly every sector: from acoustical devices to construction equipment, from dairy and livestock production to energy distribution, and many more. ANSI is also actively engaged in accrediting programs that assess conformance to standards—including globally recognized cross-sector programs such as the ISO 9000 (quality) and ISO 14000 (environmental) management systems. Their mission is "to enhance both the global competitiveness of U.S. business and the U.S. quality of life by promoting and facilitating voluntary consensus standards and conformity assessment systems, and safeguarding their integrity."

American Petroleum Institute (API) National trade association that represents all aspects of America's oil and natural gas industry. Corporate members include the largest major oil company to the smallest of independents and come from all segments of the industry. They are producers, refiners, suppliers, pipeline operators, and marine transporters, as well as service and supply companies that support all segments of the industry.

American Society of Mechanical Engineers (ASME) A not-for-profit professional organization that enables collaboration, knowledge sharing, and skill development across all engineering disciplines, while promoting the vital role of the engineer in society. ASME codes and standards, publications, conferences, continuing education, and professional development programs provide a foundation for advancing technical knowledge and a safer world.

API (American Petroleum Institute) Gravity The density measure used for petroleum liquids. API gravity is inversely related to specific gravity—the higher the API gravity, the lower the specific gravity. Temperature will affect API gravity and it should always be corrected to 60°F (16°C). API gravity can be calculated using the formula: API Gravity = 141.5/Specific Gravity – 131.5.

Aqueous Film Forming Foam (AFFF) Synthetic Class B firefighting foam consisting of fluorochemical and hydrocarbon surfactants combined with high boiling point solvents and water. AFFF film formation is dependent upon the difference in surface tension between the fuel and the firefighting foam. The fluorochemical surfactants reduce the surface tension of water to a degree less than the surface tension of the hydrocarbon so that a thin aqueous film can spread across the fuel. AFFF is not an effective extinguishing agent for natural gas since it is lighter than air gas.

Aromatic Hydrocarbon A hydrocarbon that contains the benzene "ring" which is formed by six carbon atoms and contains resonant bonds. Examples include benzene ($C_6H_6$) and toluene ($C_7H_8$).

Atmospheric Tank A storage tank that has been designed to operate at pressures from atmospheric through 0.5 psig (760–786 mmHg) measured at the top of the tank.

Aviation Gasoline (AVGAS) A gasoline fuel for reciprocating piston engine aircraft. AVGAS is very volatile and is extremely flammable at normal temperatures. AVGAS grades are defined primarily by their octane rating—the lean mixture rating and the rich mixture rating. For example, AVGAS 100/130 has a lean mixture performance rating of 100 and a rich mixture rating of 130.

Barrel A unit of measurement equal to 42 U.S. standard gallons.

Boiling Liquid Expanding Vapor Explosion (BLEVE) A container failure with a release of energy, often rapidly and violently, which is accompanied by a release of gas to the atmosphere and propulsion of the container or container pieces due to an overpressure rupture.

Boiling Point The temperature at which a liquid changes its phase to a vapor or gas; also, the temperature where the vapor pressure of the liquid equals atmospheric pressure. Significant property for evaluating the flammability of a liquid, as flash point, boiling point, and vapor pressure are directly related. A liquid with a low flash point will also have a low boiling point, which translates into a large amount of vapors being given off.

Boilover Violent ejection of flammable liquid from its container caused by the vaporization of water beneath the body of liquid. It may occur after a lengthy burning period of products such as crude oil when the heat wave has passed down through the liquid and reaches the water bottom in a storage tank. It will not occur to any significant extent with water-soluble liquids or light hydrocarbon products such as gasoline.

British Thermal Unit (BTU) The amount of heat energy needed to raise the temperature of one pound of water by one degree Fahrenheit.

Bulk Storage Plant or Terminal The portion of a facility where liquids are received by tank vessel, pipelines, tank car, or tank vehicle; stored or blended in bulk; and then distributed to the end user by one or more of the transportation modes.

Burnback Resistance The ability of a foam blanket to resist direct flame impingement such as would be evident in a partially extinguished petroleum fire.

Chemical Protective Clothing (CPC) Single or multipiece garment constructed of chemical protective clothing materials designed and configured to protect the wearer's torso, head, arms, legs, hands, and feet; can be constructed as a single or multipiece garment. The garment may completely enclose the wearer either by itself or in combination with the wearer's respiratory protection, attached or detachable hood, gloves, and boots.

Chemical Transportation Emergency Center (CHEMTREC™) The Chemical Transportation Center, operated by the American Chemistry Council (ACC), can provide information and technical assistance to emergency responders. (Phone number: 1-800-424-9300)

Class B Foam A firefighting foam designed to extinguish Class B fuels. Class B fuels can be subdivided into two subclasses: (1) hydrocarbons such as gasoline, kerosene, and fuel oil that will not mix with water, and (2) polar solvents such as alcohols, ketones, and ethers which will mix with water.

Chemically, Class B foams can be divided into two general categories: synthetic based or protein based. Synthetic foams are basically super soap with fire performance additives. They include high expansion foam, aqueous film forming foam (AFFF), and alcohol-resistant aqueous film forming foam (AR-AFFF). In general, synthetic foams flow more freely and provide quick knockdown with limited postfire security. Protein foams use natural protein foamers instead of a synthetic soap, and similar fire performance components are added. Protein type foams include regular protein foam (P), fluoroprotein foam (FP), alcohol-resistant fluoroprotein foam (AR-FP), film forming fluoroprotein (FFFP), and alcohol-resistant film forming fluoroprotein (AR-FFFP). In general, protein-based foams spread slightly slower than synthetic, but they produce a more heat-resistant, longer lasting foam blanket. [*Source*: ANGUS FIRE.]

Classes (Electrical) As used in NFPA 70—The National Electric Code, used to describe the type of flammable materials that produce the hazardous atmosphere. There are three classes of electrical locations:

Class I Locations—Flammable gases or vapors may be present in quantities sufficient to produce explosive or ignitable mixtures.

Class II Locations—Concentrations of combustible dusts may be present (e.g., coal or grain dust).

Class III Locations—Areas concerned with the presence of easily ignitable fibers or flyings (e.g., cotton milling).

Clean Air Act (CAA) Federal legislation which resulted in EPA regulations and standards governing airborne emissions, ambient air quality, and risk management programs.

Clean Water Act (CWA) Federal legislation which resulted in EPA and state regulations and standards governing drinking water quality, pollution control, and enforcement. The Oil Pollution Act (OPA) amended the CWA and authorized regulations pertaining to oil spill preparedness, planning, response, and cleanup.

Closed Container A container sealed by means of a lid or other device so that neither liquid nor vapor will escape from it at ordinary temperatures.

Code of Federal Regulations (CFR) A collection of regulations established by federal law. Contact with the agency that issues the regulation is recommended for both details and interpretation.

Combustible Liquid A liquid having a flash point at or above 100°F (37.8°C). Combustible liquids are subdivided as follows:

Class II liquids shall include those having flash points at or above 100°F (37.8°C) and below 140°F (60°C).

Class IIIA liquids shall include those having flash points at or above 140°F (60°C) and below 200°F (93°C).

Class IIIB liquids shall include those having flash points at or above 200°F (93°C).

Comprehensive Environmental Response, Compensation and Liability Act (CERCLA) Known as CERCLA or SUPERFUND, it addresses hazardous substance releases into the environment and the cleanup of inactive hazardous waste sites. It also requires those who release hazardous substances, as defined by the Environmental Protection Agency (EPA), above certain levels (known as "reportable quantities") to notify the National Response Center.

Confined Space A space that (1) is large enough and so configured that an employee can bodily enter and perform assigned work, (2) has limited or restricted means for entry or exit (e.g., tanks, vessels, silos, storage bins, hoppers, vaults, and pits are spaces that may have limited means of entry), and (3) is not designed for continuous employee occupancy.

Confinement Procedures taken to keep a material in a defined or localized area once released.

Container Any vessel of 119 gallons (450 L) or less capacity used for transporting or storing liquids.

Containment Actions necessary to keep a material in its container (e.g., stop a release of the material or reduce the amount being released).

Control The offensive or defensive procedures, techniques, and methods used in the mitigation of a hazardous materials incident, including containment, confinement, and extinguishment.

Control Zones The designation of areas at a hazardous materials incident based upon safety and the degree of hazard. Many terms are used to describe these control zones; however, for the purposes of this text, these zones are defined as the hot, warm, and cold zones. May also be referred to as hazard control zones.

Controlled Burn Defensive or nonintervention tactical objective by which a fire is allowed to burn with no effort to extinguish the fire. In some situations, extinguishing a fire will result in large volumes of contaminated runoff or threaten the safety of emergency responders. Consult with the appropriate environmental agencies when using this method.

Crude Oil A mixture of oil, gas, water, and other impurities, such as metallic compounds and sulfur. Its color can range from yellow to black. This mixture includes various petroleum fractions with a wide range of boiling points. The exact composition of crude oil varies depending upon from where in the world the crude oil was produced.

Defensive Tactics These are less aggressive spill and fire control tactics where certain areas may be "conceded" to the emergency, with response efforts directed toward limiting the overall size or spread of the problem. Examples include isolating the pipeline by closing remote valves, shutting down pumps, constructing dikes, and exposure protection.

Degree of Solubility An indication of how well one material can mix with another. Also known as miscibility.

Negligible: less than 0.1%

Slight: 0.1% to 1.0%

Moderate: 1% to 10%

Appreciable: greater than 10%

Complete: soluble at all proportions

Dilution Application of water to water-miscible flammable liquids to reduce to safe levels the hazard they represent. It can increase the total volume of liquid which will have to be disposed of. In decon applications, it is the use of water to flush a hazmat from protective clothing and equipment, and the most common method of decon.

Distillate Fuel Oils Include both diesel fuel and fuel oil. Diesel fuel is a light hydrocarbon mixture for diesel engines, similar to furnace fuel oil, but with a slightly lower boiling point. Refined fuel oil comes in two grades. No. 1 distillate, such as kerosene, is a light fuel. No. 2 fuel oil is a distillate fuel oil prepared for use as a fuel for atomizing-type burners or for smaller industrial burner units.

Divisions (Electrical) As used in NFPA 70—The National Electric Code, describe the types of location that may generate or release a flammable material. There are two divisions:

***Division I***—Location where the vapors, dusts, or fibers are continuously generated and released. The only element necessary for a hazardous situation is a source of ignition.

*Division II*—Location where the vapors, dusts, or fibers are typically confined, but can be generated and released as a result of an emergency or a failure in the containment system.

Divisions (Incident Command System) As used within the incident command system, divisions are the organizational level having responsibility for operations within a defined geographic area. Divisions are under the direction of a supervisor

Eductor A device that uses the Venturi principle to introduce a proportionate quantity of foam concentrate into a water stream. The pressure at the throat is below atmospheric pressure and will draw in liquid from atmospheric storage containers or tanks.

Emergency Response Personnel Those assigned to organizations that have the responsibility for responding to different types of emergency situations.

Emergency Response Plan A plan that establishes guidelines for handling flammable liquid incidents as required by regulations such as SARA, Title III, and HAZWOPER (29 CFR 1910.120).

Emergency Response Team (ERT) Crews of specially trained personnel used within industrial facilities for the control and mitigation of emergency situations. May consist of both shift personnel with ERT responsibilities as part of their job assignment (e.g., plant operators) or volunteer members. ERTs may be responsible for fire, hazmat, oil spill, medical, and technical rescue emergencies, depending on the size and operation of the facility.

Environmental Protection Agency (EPA) The purpose of the EPA is to protect and enhance our environment today and for future generations to the fullest extent possible under the laws enacted by Congress. The EPA's mission is to control and abate pollution in the areas of water, air, solid waste, pesticides, noise, and radiation. EPA's mandate is to mount an integrated, coordinated attack on environmental pollution in cooperation with state and local governments.

Expansion Ratio The amount of gas produced by the evaporation of one volume of liquid at a given temperature. Significant property when evaluating liquid and vapor releases of liquefied gases and cryogenic materials: the greater the expansion ratio, the more gas that is produced and the larger the hazard area.

Explosion-Proof Construction Encases the electrical equipment in a rigidly built container so that (1) it withstands the internal explosion of a flammable mixture, and (2) prevents propagation to the surrounding flammable atmosphere. Used in Class I, Division 1 atmospheres at fixed installations. Being explosion-proof does not mean the enclosure is gas tight.

Failure of Container Attachments Attachments which open up or break off the container, such as safety relief valves, frangible discs, fusible plugs, discharge valves, or other related appliances.

Film-Forming Fluoroprotein (FFFP) Concentrate Composed of a combination of protein and film-forming surfactants. The foam formed acts as a barrier to exclude air or oxygen and develops an aqueous film on some fuels that suppresses the evolution of fuel vapors. Concentrates are diluted with water to a 3% or 6% solution by volume, depending on the type of concentrate.

Fire Entry Suits Suits which offer complete, effective protection for short duration entry into a total flame environment. Designed to withstand exposures to radiant heat levels up to 2,000°F. Entry suits consist of a coat, pants, and separate hood assembly. They are constructed of several layers of flame-retardant materials, with the outer layer often aluminized.

Fire Point Minimum temperature at which a liquid gives off sufficient vapors that will ignite and sustain combustion. It is typically several degrees higher than the flash point. In assessing the risk posed by a flammable liquids release, greater emphasis is placed upon the flash point, since it is a lower temperature and sustained combustion is not necessary for significant injuries or damage to occur. For liquids with very low flash points, for example, gasoline (flash point −40°F), there is no practical difference between the flash point and the fire point.

First Responder The first trained person(s) to arrive at the scene of a hazardous materials incident. May be from the public or private sector of emergency services.

First Responder, Awareness Level Individuals who are likely to witness or discover a hazardous substance release who have been trained to initiate an emergency response sequence by notifying the proper authorities of the release. They would take no further action beyond notifying the authorities of the release.

First Responder, Operations Level Individuals who respond to releases or potential releases of hazardous substances as part of the initial response to the site for the purpose of protecting nearby persons, property, or the environment from the effects of the release. They are trained to respond in a defensive fashion without actually trying to stop the release. Their function is to contain the release from a safe distance, keep it from spreading, and prevent exposures.

Flammable Atmospheres The explosive range for most common hydrocarbon liquids, such as gasoline, heating oils, diesel fuels, and aviation gasoline, typically is from 1% to 10% vapor-to-air mixture; however, the explosive range for oxygenated hydrocarbons such as alcohols and glycols is wider.

Flammable (Explosive) Range The range of gas or vapor concentration (percentage by volume in air) that will burn or explode if an ignition source is present. Limiting concentrations are commonly called the "lower flammable (explosive) limit" and the "upper flammable (explosive) limit." Below the lower flammable limit, the mixture is too lean to burn; above the upper flammable limit, the mixture is too rich to burn. If the gas or vapor is released into an oxygen-enriched atmosphere, the flammable range will expand. Likewise, if the gas or vapor is released into an oxygen-deficient atmosphere, the flammable range will contract.

Flammable Liquid A liquid having a flash point below 100°F (37.8°C) and having a vapor pressure not exceeding 40 psig (absolute) (2,068 mmHg) at 100°F (37.8°C) shall be known as a Class I liquid.

- Class I liquids shall be subdivided as follows:

  Class IA shall include those having flash points below 73°F (22.8°C) and having a boiling point below 100°F (37.8°C).

  Class IB shall include those having flash points below 73°F (22.8°C) and having a boiling point at or above 100°F (37.8°C).

  Class IC shall include those having flash points at or above 73°F (22.8°C) and below 100°F (37.8°C). [*Source*: NFPA 30]

Flaring Controlled burning of a high vapor pressure liquid or compressed gas in order to reduce or control the pressure and/or dispose of the product.

Flash Point Minimum temperature at which a liquid gives off enough vapors that will ignite and flashover but will not continue to burn without the addition of more heat. Significant in determining the temperature at which the vapors from a flammable liquid are readily available and may ignite.

Fluoroprotein (FP) Foam Concentrate A foam concentrate with a protein base and a synthetic fluorinated surfactant additive. In addition to an air-excluding foam blanket, it may also deposit a vaporization-preventing film on the surface of a liquid fuel. Concentrates are diluted with water to a 3% or 6% solution by volume, depending on the type of concentrate.

Foam A firefighting foam is simply a stable mass of small air-filled bubbles that have a lower density than oil, gasoline, or water. Foam is made up of three ingredients: water, foam concentrate, and air. When mixed in the correct proportions, these three ingredients form a homogeneous foam blanket.

Foam Application Rate Measure of the quantity of foam applied per unit of time per unit of area. It is usually based on the amount of foam solution (in gallons or liters) per unit of time (in minutes) per unit of area (in square feet or square meters); for example, gallons per minute per square foot.

Foam Concentrate The foaming agent for mixing with the appropriate amounts of water and air to produce finished foam.

Foam Expansion Value The ratio of final foam volume to original foam solution volume before adding air. It is also the numerical value of the reciprocal of the specific gravity of the foam.

Foam Maker A device designed to introduce air into a pressurized foam solution stream.

Foam Quality A measure of a foam's physical characteristics, expressed as the foam's 15% drain time, expansion value, and burnback resistance.

Foam Stability The relative ability of a foam to withstand spontaneous collapse or breakdown from external causes such as heat or chemical reaction.

Frothover Can occur when water already present inside a tank comes in contact with a hot viscous oil which is being loaded. A typical example occurs when hot asphalt is loaded into a tank car that contains some water. The asphalt is initially cooled by contact with the cold metal, and there is no reaction. However, the water can later become superheated, eventually start to boil, and cause the asphalt to overflow the tank car. A similar situation can arise when a tank used to store slops or residuum at temperatures below 200°F (93°C) and that contains a water bottom or oil-in-water (wet) emulsion receives a substantial addition of hot residuum at a temperature well above 212°F (100°C). After sufficient time has passed for the effect of the hot oil to reach the water in the tank, a prolonged boiling action can occur. This boiling action can rupture the tank roof and spread superheated froth over a wide area.

Full Protective Clothing Protective clothing worn primarily by firefighters which includes helmet, fire retardant hood, coat, pants, boots, gloves, PASS device, and self-contained breathing apparatus designed for structural firefighting. It does not provide specialized chemical splash or vapor protection.

Grounding A method of controlling ignition hazards from static electricity. The process of connecting one or more conductive objects to the ground; to minimize potential differences between objects and the ground. In other words, the objects are at zero potential. Also referred to as "earthing."

Groups (Electrical) As used in NFPA 70—The National Electric Code, are products within a class. Class I is divided into four groups (Groups A–D) on the basis of similar flammability characteristics. Class II is divided into three groups (Groups E–G). There are no groups for Class III materials.

Groups (Incident Command System) As used within the incident command system, are the organizational levels responsible for specified functional assignments at an incident (e.g., hazmat group). Groups are under the direction of a supervisor and may move between divisions at an incident.

Hazard A danger or peril. In hazmat operations, usually refers to the physical or chemical properties of a material.

Hazard and Risk Evaluation Evaluation of hazard information and the assessment of the relative risks of a hazmat incident. Evaluation process leads to the development of an incident action plan.

Hazard Control Zones The designation of areas at a hazardous materials incident based upon safety and the degree of hazard. Many terms are used to describe these hazard control zones; however, for the purposes of this text, these zones are defined as the hot, warm, and cold zones. May also be referred to as control zones.

Hazardous Materials (1) Any substance or material in any form or quantity that poses an unreasonable risk to safety and health and property when transported in commerce. (*Source*: U.S. Department of Transportation [DOT], 49 Code of Federal Regulations (CFR) 171) (2) Any substance that jumps out of its container when something goes wrong and hurts or harms the things it touches. (*Source*: Ludwig Benner, Jr.)

Hazardous Materials Specialists Individuals who respond and provide support to hazardous materials technicians. While their duties parallel those of the technician, they require a more detailed or specific knowledge of the various substances they may be called upon to contain. Would also act as a liaison with federal, state, local, and other governmental authorities in regard to site activities.

Hazardous Materials Technicians Individuals who respond to releases or potential releases of hazardous materials for the purposes of stopping the leak. They generally assume a more aggressive role in that they are able to approach the point of a release in order to plug, patch, or otherwise stop the release of a hazardous substance.

Hazardous Substance Any substance designated under the Clean Water Act and the Comprehensive Environmental Response, Compensation and Liability Act (CERCLA) as posing a threat to waterways and the environment when released (U.S. Environmental Protection Agency, 40 CFR 302). Hazardous substances as used within OSHA 1910.120 refers to every chemical regulated by EPA as a hazardous substance and by DOT as a hazardous material.

Hazmat Abbreviation used for hazardous materials.

High-Backpressure Foam Maker An inline aspirator used to deliver foam under pressure. Air is supplied directly to the foam solution through a Venturi action, which results in a low air content but in homogeneous and stable foam.

High Level Alarms (HLA) Alerting devices designed to minimize the risk of a tank overfill prevention system when storage tanks are receiving product from a pipeline and are in danger of being overfilled. *API Standard 2350—Overfill Protection for Storage Tanks in Petroleum Facilities, Fourth Edition* identifies three levels of concern (LOC) for tank overfill prevention. These include:

1. Maximum Working Level (MWL)—An operational level that is the highest product level to which the tank may routinely be filled during normal operations.
2. Level Alarm High-High (LAHH)—An alarm is generated when the product level reaches the high-high tank level. The alarm requires immediate action.
3. Critical High (CH) Level—The highest level in the tank that product can reach without detrimental impacts (i.e., product overflow or tank damage).

Note: Historically, tank overfills have been a leading cause of serious incidents in the petroleum and related process industries. Tank overfill was the cause of both the 2005 Buncefield, England, and the 2009 Bayamon, Puerto Rico, fires and vapor cloud explosions.

High Temperature Protective Clothing Protective clothing designed to protect the wearer against short-term high temperature exposures. Includes both proximity suits and fire entry suits. This type of clothing is usually of limited use in dealing with chemical exposures.

Hot Tapping A technique for welding on and cutting holes through liquid and/or compressed gas vessels and piping while in service for the purposes of relieving the internal pressure and/or removing the product.

Hot Zone An area immediately surrounding a hazardous materials incident, which extends far enough to prevent adverse effects from hazardous materials releases to personnel outside the zone. This zone is also referred to as the "exclusion zone," the "red zone," and the "restricted zone" in other documents. Law enforcement personnel may also refer to this as the inner perimeter.

Hydrocarbons Compounds primarily made up of hydrogen and carbon. Examples include LPG, gasoline, and fuel oils.

Hydrogen Sulfide ($H_2S$) A colorless, very poisonous, flammable gas with the characteristic foul odor of rotten eggs. It often results from the bacterial breakdown of organic matter in the absence of oxygen, such as in swamps and sewers. It also occurs in volcanic gases, natural gas, and some well waters.

Ignition (Autoignition) Temperature Minimum temperature required to ignite gas or vapor without a spark or flame being present. Significant in evaluating the ease at which a flammable material may ignite due to contact with its environment.

Immediately Dangerous to Life or Health (IDLH) An atmospheric concentration of any toxic, corrosive, or asphyxiant substance that poses an immediate threat to life or would cause irreversible or delayed adverse health effects or would interfere with an individual's ability to escape from a dangerous atmosphere.

Incident (1) The release or potential release of a hazardous material from its container into the environment. (2) An occurrence or event, either natural or humanmade, which requires action by emergency response personnel to prevent or minimize loss of life or damage to property and/or natural resources.

Incident Action Plan The strategic goals, tactical objectives, and support requirements for the incident. All incidents require an action plan. For simple incidents (Level I) the action plan is not usually in written form. Large or complex incidents (Level II or III) will require that the action plan be documented in writing.

Incident Commander (IC) The individual responsible for establishing and managing the overall incident action plan (IAP). This process includes developing an effective organizational structure, developing an incident strategy and tactical action plan, allocating resources, making appropriate assignments, managing information, and continually attempting to achieve the basic command goals. The IC is in charge of the incident site. (May also be referred to as the On-Scene Incident Commander as defined in 29 CFR. 1910.120.)

Incident Command Post (ICP) The onscene location where the IC develops goals and objectives, communicates with subordinates, and coordinates activities between various agencies and organizations. The ICP is the "field office" for onscene response operations, and requires access to communications, information, and both technical and administrative support.

Incident Command System (ICS) An organized system of roles, responsibilities, and standard operating procedures used to manage and direct emergency operations. May also be referred to the incident management system (IMS).

Inert Gas A nonreactive, nonflammable, noncorrosive gas, such as argon, helium, krypton, neon, nitrogen, and xenon. [*Source*: NFPA 1]

Intrinsically Safe Construction Equipment or wiring that is incapable of releasing sufficient electrical energy under both normal and abnormal conditions to cause the ignition of a flammable mixture. Commonly used in portable direct-reading instruments for operations in Class I, Division 2 hazardous locations.

Isolation Perimeter The designated crowd control line surrounding the hazard control zones. The isolation perimeter is always the line between the general public and the cold zone. Law enforcement personnel may also refer to this as the outer perimeter.

Jet Fuel A highly refined kerosene petroleum distillate. There are two main grades of jet fuel used in commercial aviation—Jet A and Jet A-1. Both are kerosene distillates. Jet A is found in the United States and has a flash point above 100°F (38°C). Jet A-1 is a similar kerosene fuel that is used for civil commercial aviation internationally. Jet B is a blend of kerosene and gasoline, and may be found in extremely cold climates where cold weather performance is critical (e.g., Canada).

Leak The uncontrolled release of a hazardous material which could pose a threat to health, safety, and/or the environment.

Local Emergency Planning Committee (LEPC) A committee appointed by a State Emergency Response Commission, as required by SARA Title III, to formulate a comprehensive emergency plan for its region.

Lower Explosive Limit (LEL) The lowest concentration at which a gas or vapor is flammable or explosive at ambient conditions.

Low-Pressure Tank A storage tank designed to withstand an internal pressure above 0.5 psig (3.5 kPa) but not more than 15 psig (103.4 kPa) measured at the top of the tank.

Minimum Application Rate Rate of foam application that is sufficient to cause extinguishment and demonstrate satisfactory stability and resistance to burnback (see NFPA 11).

Mitigation Any offensive or defensive action to contain, control, reduce, or eliminate the harmful effects of a hazardous materials release.

Monitoring The act of systematically checking to determine contaminant levels and atmospheric conditions.

Monitoring Instruments They include, but are not limited to, monitoring and detection instruments used to detect the presence and/or concentration of contaminants within an environment. They include Combustible Gas Indicators (CGIs), Oxygen Monitors, Colorimetric Indicator Tubes, Specific Chemical Monitors, Flame Ionization Detectors (FIDs), Gas Chromatographs, Photoionization Detectors (PIDs), Radiation Monitors, Radiation Dosimeter Detectors, Corrosivity (pH) Detectors, and Indicator Papers.

National Fire Protection Association (NFPA) An international voluntary membership organization to promote improved fire protection and prevention, establish safeguards against loss of life and property by fire, and write and publish national voluntary consensus standards (e.g., *NFPA 472—Professional Competence of Responders to Hazardous Materials Incidents*).

National Incident Management System (NIMS) A standardized systems approach to incident management that consists of five major subdivisions collectively providing a total systems approach to all-risk incident management.

National Institute for Occupational Safety and Health (NIOSH) A federal agency which, among other activities, tests and certifies respiratory protective devices and air sampling detector tubes, and recommends occupational exposure limits for various substances.

National Response Center (NRC) Communications center operated by the U.S. Coast Guard in Washington, D.C. It provides information on suggested technical emergency actions, and is the federal spill notification point. The NRC must be notified within 24 hours of any spill of a reportable quantity of a hazardous substance by the spiller. Can be contacted at (800) 424-8802.

Nonintervention Tactics Essentially "no action." It is useful at certain fire emergencies where the potential costs of action far exceed any benefits (e.g., BLEVE scenario).

Occupational Safety and Health Administration (OSHA) Component of the U.S. Department of Labor; an agency with safety and health regulatory and enforcement authorities for most U.S. industries, businesses, and states.

Offensive Tactics Aggressive leak, spill, and fire control tactics designed to quickly control or mitigate the problem. Although increasing risks to emergency responders, offensive tactics may be justified if rescue operations can be quickly achieved, if the spill can be rapidly confined or contained, or if the fire can be quickly extinguished.

Oil Pollution Act (OPA) Amended the federal Water Pollution Act. OPA's scope covers both facilities and carriers of oil and related liquid products, including deepwater marine terminals, marine vessels, pipelines, and railcars. Requirements include the development of emergency response plans, training and exercises, and verification of spill resources and contractor capabilities.

On-Scene Coordinator (OSC) The federal official predesignated by EPA or the USCG to coordinate and direct federal responses and removals under the National Contingency Plan.

On-Scene Incident Commander *See* Incident Commander.

Oxygen-Deficient Atmosphere An atmosphere that contains an oxygen content less than 19.5% by volume at sea level.

Personal Protective Equipment (PPE) Equipment provided to shield or isolate a person from the chemical, physical, and thermal hazards that may be encountered at a hazardous materials incident. Adequate personal protective equipment should protect the respiratory system, skin, eyes, face, hands, feet, head, body, and hearing. Personal protective equipment includes personal protective clothing, self-contained positive pressure breathing apparatus, and air purifying respirators.

Polar Solvent A liquid whose molecules possess a permanent electric moment. Examples are amines, ethers, alcohols, esters, aldehydes, and ketones. In firefighting, any flammable liquid which destroys regular foam is generally referred to as a polar solvent (or is water miscible).

Polymeric Membrane A thin, durable, cohesive skin formed on a polar solvent fuel surface, protecting the foam bubbles from destruction by the fuel; a precipitation that occurs when a polar solvent foam comes in contact with hydrophilic fuels such as isopropanol, ethanol, and other polar solvents.

Pressure Relief Valve (PRV) A type of pressure relief device designed to open and close in order to maintain internal pressure within a container or pipeline and minimize the risk of container failure.

Process Safety Management (PSM) The application of management principles, methods, and practices to prevent and control releases of hazardous chemicals or energy. Focus of both OSHA 1910.119—Process Safety Management of Highly Hazardous Chemicals, Explosives and Blasting Agents and EPA Part 68—Risk Management Programs for Chemical Accidental Release Prevention.

Proportioner The device where foam liquid and water are mixed to form foam solution.

Protective Clothing Equipment designed to protect the wearer from heat and/or hazardous materials contacting the skin or eyes. Protective clothing is divided into four types:

- Structural fire fighting clothing
- Liquid chemical splash protective clothing
- Chemical vapor protective clothing
- High temperature protective clothing

Proximity Suits Designed for exposures of short duration and close proximity to flame and radiant heat, such as in aircraft rescue firefighting (ARFF) operations. The outer shell is a highly reflective, aluminized fabric over an inner shell of a flame-retardant fabric such as Kevlar™ or Kevlar™/PBI™ blends. These ensembles are not designed to offer any substantial chemical protection.

Purging The process of supplying an enclosure with a protective gas at a sufficient flow and positive pressure to reduce the concentration of any flammable gas or vapor initially present to an acceptable level. Used in Class I, Division 1 atmospheres at fixed installations. [*Source*: NFPA 496]

Refined Products Liquid petroleum products produced through the refining process. Examples include gasoline, aviation gasoline, jet fuels, and distillates such as home heating and diesel fuels.

Reid Vapor Pressure (RVP) The rate at which vapors are given off from a petroleum liquid. It is commonly used within the pipeline industry when discussing winter and summer gasoline blends.

Residual Pressure The pressure existing in a water main or hoseline at a specified flow (as opposed to static pressure).

Resource Conservation and Recovery Act (RCRA) Law that establishes the regulatory framework for the proper management and disposal of all hazardous wastes, including treatment, storage, and disposal facilities. It also establishes installation, leak prevention, and notification requirements for underground storage tanks.

Respiratory Protection Equipment designed to protect the wearer from the inhalation of contaminants. Respiratory protection includes positive-pressure self-contained breathing apparatus (SCBA), positive-pressure airline respirators (SARs), and air purifying respirators.

Risk The probability of suffering a harm or loss. Risks are variable and change with every incident.

Risk Analysis A process to analyze the probability that harm may occur to life, property, and the environment and to note the risks to be taken to identify the incident objectives.

Risk-Based Response Process Systematic process by which responders analyze a problem involving hazardous materials/weapons of mass destruction (WMDs), assess the hazards, evaluate the potential consequences, and determine appropriate response actions based on facts, science, and the circumstances of the incident.

Sanitary Sewer A "closed" sewer system which carries wastewater from individual homes, together with minor quantities of storm water, surface water, and ground water that are not admitted intentionally. May also collect wastewater from industrial and commercial businesses. The collection and pumping system will transport the wastewater to a treatment plant, where it is processed.

Saturated Hydrocarbons A hydrocarbon possessing only single covalent bonds. All of the carbon atoms are saturated with hydrogen. Examples include methane (CH4), propane (C3H8), and butane (C4H10).

Secondary Containment Tank A tank having an inner and an outer wall with an interstitial space (annulus) between the walls and having means for monitoring the interstitial space for a leak.

Self-Contained Breathing Apparatus (SCBA) A positive pressure, self-contained breathing apparatus (SCBA) or combination SCBA/supplied air breathing apparatus certified by the National Institute for Occupational Safety and Health (NIOSH) and the Mine Safety and Health Administration (MSHA), or the appropriate approval agency for use in atmospheres that are immediately dangerous to life or health (IDLH).

Site Management and Control The management and control of the physical site of a hazmat incident. Includes initially establishing command, approach and positioning, staging, initial perimeter and hazard control zones, and public protective actions.

Size-Up The rapid, yet deliberate consideration of all critical scene factors.

Slopover Can result when a water stream is applied to the hot surface of a burning oil, provided the oil is viscous and its temperature exceeds the boiling point of water. It can also occur when the heat wave contacts a small amount of stratified water within a crude oil. As with a boilover, when the heat wave contacts the water, the water converts to steam and causes the product to "slopover" the top of the tank. Slopovers can range from a quietlike boiling of the product over the tank to a large explosion of burning slop.

Solubility The ability of a solid, liquid, gas, or vapor to dissolve in water or other specified medium. The ability of one material to blend uniformly with another, as in a solid in liquid, liquid in liquid, gas in liquid, or gas in gas. Significant property in evaluating the selection of control and extinguishing agents, including the use of water and firefighting foams.

Specific Gravity The weight of the material as compared with the weight of an equal volume of water. If the specific gravity is less than 1, the material is lighter than water and will float. If the specific gravity is greater than 1, the material is heavier than water and will sink. Most insoluble hydrocarbons are lighter than water and will float on the surface. Significant property for determining spill control and cleanup procedures for waterborne releases.

Spill The release of a liquid, powder, or solid hazardous material in a manner that poses a threat to air, water, ground, and the environment.

Static Pressure The pressure existing in a water or hoseline with no water flowing. This can be considerably higher than the residual pressure. *See* Residual Pressure.

Steel Tank Institute (STI). A not-for-profit organization that works with tank manufacturers, users, regulatory authorities, and consultants to promulgate standards for the design, construction, and installation of aboveground and underground tanks used for the storage of flammable and combustible liquids. The STI publishes consensus standards that guide tank manufacturers and code enforcement officials.

Storage Tank Any vessel having a liquid capacity that exceeds 60 gallons (227 L), is intended for fixed installation, and is not used for processing.

Storm Sewer An "open" system which collects storm water, surface water, and ground water from throughout an area, but excludes domestic wastewater and industrial wastes. A storm sewer may dump runoff directly into a retention area, which is normally dry, or into a stream, river, or waterway without treatment.

Strategic Goals The overall plan that will be used to control an incident. Strategic goals are broad in nature and are achieved by the completion of tactical objectives. Examples include rescue, spill control, leak control, and recovery.

Structural Firefighting Protective Clothing Protective clothing normally worn by firefighters during structural firefighting operations. It includes a helmet, coat, pants, boots, gloves, PASS device, and a hood to cover parts of the head not protected by the helmet. Structural firefighting clothing provides limited protection from heat, but it may not provide adequate protection from harmful liquids, gases, vapors, or dusts encountered during hazmat incidents. May also be referred to as turnout or bunker clothing.

Submergence Plunging of foam beneath the surface of burning liquid, resulting in a partial breakdown of the foam structure and coating of the foam with the burning liquid.

Subsurface Injection A method of fighting hydrocarbon tank fires in which foam is fed into the tank at a point below the surface of the burning fuel. The foam rises to the surface and blankets the fuel vapor at the surface.

Superfund Amendments & Reauthorization Act (SARA) Created for the purpose of establishing federal statutes for right-to-know standards and emergency response to hazardous materials incidents, reauthorized the federal Superfund program, and mandated states to implement equivalent regulations/requirements.

Tactical Objectives The specific operations that must be accomplished to achieve strategic goals. Tactical objectives must be both specific and measurable.

Task. The specific activities that accomplish a tactical objective.

Thermal Stress Hazmat container stress generally indicated by temperature extremes, both hot and cold. Examples include fire, sparks, friction or electricity, and ambient temperature changes. Extreme or intense cold, such as that found with cryogenic materials, may also act as a stressor. Clues of thermal stress include the operation of pressure relief devices and/or the bulging of containers.

Threshold The point where a physiological or toxicological effect begins to be produced by the smallest degree of stimulation.

Threshold Limit Value/Time Weighted Average (TLV/TWA) The airborne concentration of a material to which an average, healthy person may be exposed repeatedly for 8 hours each day, 40 hours per week, without suffering adverse effects. The young, old, ill, and naturally susceptible will have lower tolerances and will need to take additional precautions. TLVs are based upon current available information and are adjusted on an annual basis by organizations such as the American Conference of Governmental Industrial Hygienists (ACGIH). As TLVs are time weighted averages over an 8-hour exposure, they are difficult to correlate to emergency response operations. The lower the value, the more toxic the substance.

Topside Application A method of foam discharge in which the foam is applied to the surface of the burning fuel.

Toxic Products of Combustion The toxic by-products of the combustion process. Depending upon the materials burning, higher levels of personal protective clothing and equipment may be required.

Toxicity The ability of a substance to cause injury to a biological tissue. Refers to the ability of a chemical to harm the body once contact has occurred.

Transfer The process of physically moving a liquid, gas, or some form of solid either manually, by pump, or by pressure transfer, from a leaking or damaged container. The transfer pump, hoses, fittings, and container must be compatible with the hazardous materials involved. When transferring flammable liquids, proper bonding and grounding concerns must be addressed.

Twenty-five percent (25%) Drain Time The time required for 25% of the liquid contained in the foam to drain.

Type I Discharge Outlet A device that conducts and delivers foam onto the burning surface of a liquid without submerging the foam or agitating the surface (e.g., a foam trough).

Type II Discharge Outlet A device that delivers foam onto the burning liquid, partially submerges the foam, and produces restricted agitation of the surface (e.g., a foam chamber).

Type III Discharge Outlet A device that delivers foam so that it falls directly onto the surface of the burning liquid in a manner that causes general agitation (e.g., lobbing with a foam nozzle).

Underwriters Laboratories (UL) An organization that helps companies demonstrate safety, confirm compliance, enhance sustainability, manage transparency, deliver quality and performance, strengthen security, protect brand reputation, build workplace excellence, and advance societal well-being. Some of the services offered by UL include inspection, advisory services, education and training, testing, auditing and analytics, certification software solutions, and marketing claim verification.

Unified Command The process of determining overall incident strategies and tactical objectives by having all agencies, organizations, or individuals who have jurisdictional responsibility, and in some cases those who have functional responsibility at the incident, participate in the decision-making process.

Unified Commanders (UCs) Command level representatives from each of the primary responding agencies who present their agency's interests as a member of a unified command team. Depending upon the scenario and incident timeline, they may be the "lead" IC or play a supporting role within the command function. The UCs manage their own agency's actions and ensure all efforts are coordinated through the unified command process.

Vapor An air dispersion of molecules in a substance that is normally a liquid or solid at standard temperature and pressure.

Vapor Density The weight of a pure vapor or gas compared with the weight of an equal volume of dry air at the same temperature and pressure. The molecular weight of air is 29. If the vapor density of a gas is less than 1, the material is lighter than air and may rise. If the vapor density is greater than 1, the material is heavier than air and will collect in low or enclosed areas. Significant property for evaluating exposures and where hazmat gas and vapor will travel.

Vapor Dispersion Use of water spray to disperse or move vapors away from certain areas or materials. Note that reducing the concentration of a material through the use of a water spray may bring the material into its flammable range.

Vapor Pressure The pressure exerted by the vapor within the container against the sides of a container. This pressure is temperature dependent; as the temperature increases, so does the vapor pressure. Consider the following three points:

1. The vapor pressure of a substance at 100°F is always higher than the vapor pressure at 68°F.
2. Vapor pressures reported in millimeters of mercury (mmHg) are usually very low pressures; 760 mmHg is equivalent to 14.7 psi or 1 atmosphere. Materials with vapor pressures greater than 760 mmHg are usually found as gases.
3. The lower the boiling point of a liquid, the greater vapor pressure at a given temperature.

Vapor Suppression Offensive control techniques used to mitigate the evolution of flammable, corrosive, or toxic vapors, and reduce the surface area exposed to the atmosphere. Includes the use of firefighting foams and hazmat vapor suppressants.

Venturi A constricted portion of a pipe or tube which increases water velocity, thus momentarily reducing its pressure. It is in this reduced pressure area that foam liquid is introduced in many types of proportioning equipment.

Violent Rupture Associated with chemical reactions having a release rate of less than 1 second (i.e., deflagration). There is no time to react in this scenario. This behavior is commonly associated with runaway cracking and overpressure of closed containers.

Viscosity Measurement of the thickness of a liquid and its ability to flow. High viscosity liquids, such as heavy oils, must first be heated to increase their fluidity. Low viscosity liquids spread more easily and increase the size of the hazard area.

Volatility The ease with which a liquid or solid can pass into the vapor state. The higher a material's volatility, the greater its rate of evaporation. Significant property in that volatile materials will readily disperse and increase the hazard area.

Warm Zone The area where personnel and equipment decontamination and hot zone support takes place. It includes control points for the access corridor and thus assists in reducing the spread of contamination. This is also referred to as "decontamination," "contamination reduction," "yellow zone," "support zone," or "limited access zone" in other documents.

Water Reactivity Ability of a material to react with water and release a flammable gas or present a health hazard.

Wind Girder A ring connected to the top of an open top floating roof tank which acts as a stiffener to strengthen the tank. It also acts as an air-foil during high winds.